More than the Soil

More than the Soil:
Rural Change in Southeast Asia

Jonathan Rigg

Routledge
Taylor & Francis Group

LONDON AND NEW YORK

First published 2001 by Pearson Education Limited

Published 2014 by Routledge
2 Park Square, Milton Park, Abingdon, Oxon OX14 4RN
711 Third Avenue, New York, NY 10017, USA

Routledge is an imprint of the Taylor & Francis Group, an informa business

ISBN 13: 978-0-582-38211-4 (pbk)

British Library Cataloguing in Publication Data
A CIP catalogue record for this book can be obtained from the British Library.

Typeset by 35 in 11/12pt Garamond

For Christopher Joseph
who introduced me, and many others, to geography

Contents

Contents

List of tables

List of figures

List of plates

List of boxes

Preface

It is February and I am writing this overlooking the harvested ricefields of the Ping Valley in Northern Thailand. To one side is the wall of a housing estate where some of Chiang Mai City's wealthy (and mobile) middle class increasingly choose to live, and where I have set up shop with my family for four months. Between the wall of the housing estate, with its manicured lawns, swimming pool, tennis courts and sculpted trees, is an area of unused land. Formerly riceland, it has been colonised by shrubs and grasses. I mean to walk along there and find out why it is idle. My guess is that it is because a wealthy Thai family have bought it for speculative reasons. Another possibility, though, is that the wall of the housing estate has disrupted the local drainage system and made wet rice cultivation a hazardous affair. The least likely scenario is that the availability of non-farm jobs in this area of Thailand, and the generally low status accorded agricultural work, has meant that there is no one to work the land.

Around 20 kilometres to the south is Tambon Thung Sadok, where I am carrying out fieldwork. The holding company that brews Mekong whiskey, as well as Carlsberg beer in Thailand has a subsidiary based in Tambon Thung Sadok that contracts farmers to grow a variety of Japanese rice for their brewery operations. Around a kilometre to the east is a canning factory which buys its fruit from orchards in the area, mostly planted on converted riceland. It also employs around 200 local people. Moving 20 kilometres further east still from Tambon Thung Sadok are the industrial estates of Lamphun that source many of their workers from rural communities in this area of Thailand.

The Regent Hotel in the Mae Sa Valley, about 15 minutes or so north of Chiang Mai, offers another insight into changing rural Thailand. At this time of year – February – most ricefields are dry. But looking out from *The Regent*, the deep-pocketed and highly cosseted guest is confronted by a view of terraces of verdant rice fields. A closer look reveals that the paddies are at different stages of maturity. The scene, it transpires, is contrived for the edification of the hotel's guests. The buffaloes that plough these few fields are almost extinct in the area, farmers having moved onto mechanical land preparation some years ago. But for *The Regent* only the dull clunk of the wooden buffalo bell and the slow turning of the moistened soil by draught animals can fulfil the bucolic vision that they have constructed for their guests. Even the spacing of the transplanted rice seedlings is wider than usual to allow the hotel's patrons to see the reflections of sky and trees in the standing water. Nature and the seasons are reworked in the interests of (largely) foreign guests and agriculture becomes just another arm of the tourist industry.

So, within eyeshot or an easy drive from where I type, a number of the themes with which this book is concerned are to be found: the infiltration of new classes with new interests into rural areas (last night residents of the housing estate were complaining of the ash in the swimming pool and on the tennis courts – produced by farmers burning their remaining rice stubble); the tensions between farm and non-farm activities; the role of rural industries in rural people's livelihoods; the penetration of industry into the countryside; the changing role of rural areas as zones of production to places of consumption; and the transformation in the hopes and aspirations of rural inhabitants.

This book is an extension and development of some of the arguments rehearsed in an earlier volume (*Southeast Asia: the Human Landscape of Modernisation and Development*, London: Routledge, 1997). That book, though, did not concern itself with rural areas per se but rather with the whole process of modernisation and development. But in researching and writing that book it became increasingly clear that some critical transformations were taking place in the Southeast Asian countryside, and in the social, spatial and economic interactions between rural and urban areas and people.

Inevitably, I have taken some liberties in constructing the argument that weaves its way through these pages. I would be the first to accept that rural Southeast Asia is a complex and diverse place where multiple processes of agrarian transition are under way. I hope that this complexity is clear from the discussion. I have not tried to produce any sort of aggregate view of the region. For the most part I have drawn on selected case studies, many qualitative, to make my case. These are scattered over several countries, regions and cultures and tell numerous stories about the processes and trajectories of rural change. Nonetheless, there is an underlying narrative which I have tried to pick out.

A second area where I feel an explanation is warranted is in terms of the balance of material between the established market economies of the region and the transitional economies (i.e. Cambodia, Laos, Myanmar and Vietnam). The great bulk of the detailed supporting evidence is drawn from the former group of countries for the simple reason that work – and in particular qualitative work – from the latter is lacking. This is particularly true of Cambodia and Myanmar. Furthermore, the processes with which the book is concerned are that much more advanced in Indonesia, Malaysia, the Philippines and Thailand. While this book is about many things, more than anything it is an attempt – to adapt historian David Landes' expression – to 'see rural people whole'.

Jonathan Rigg
Chiang Mai, Thailand
February 2000

Acknowledgements

Various people and institutions have offered their help and assistance in the writing of this book. On an individual level I am grateful to Chris Baker, Malcolm Falkus and Kevin Hewison, and to Doug Johnstone for permission to refer to their unpublished work. Giles Ungpakorn also allowed me to refer to his unpublished paper and pointed me in the right direction with regard to workers and peasants. Rodolphe de Koninck kindly provided information on developments in the Kedah villages of Matang Pinang and Paya Keladi since the publication of his monograph in 1992. Daniel Arghiros kindly looked over my attempts to represent his prose figuratively. Myo Thant at the Asian Development Bank informed me regarding labour shortages and AIDS. Chusak Wittayapak not only allowed me to use his unpublished work but was also generous with his time while I was based in Chiang Mai. Furthermore, he introduced me to my research assistant, Sakunee Nattapoolwat (Koi), with whom I happily and productively worked for three months in Sanpathong and Mae Wang. And, once again, Rachel Harrison has provided help on Thai language issues that were beyond my capabilities. Matthew Smith, my commissioning editor at Pearson Education, showed enormous faith that the book would be delivered on time and in reasonable shape. Colleagues in the Department of Geography at the University of Durham, and particularly fellow members of the Development Studies Research Group have provided me with both a stimulating environment in which to work and time to write. All these people, and more, have helped me.

At an institutional level, I am grateful to the Institute of Southeast Asian Studies (ISEAS) in Singapore for allowing me to use their excellent library facilities on more than one occasion. And during the latter stages of writing the book I was a visiting researcher at Chiang Mai University in Northern Thailand, based in the Department of Geography. I was also fortunate enough to have contact with the postgraduates attached to the new Regional Center for Social Science and Sustainable Development (RCSD). This period in Thailand was supported by a grant from the British Academy South East Asia Committee which, over the years, has been very generous in providing me with funds for fieldwork in Thailand, Laos and Indonesia. The work was also approved and supported by the National Research Council of Thailand.

Lastly, what to say about my family? Since my last book it has increased in size by one and so, from a productivity perspective and like the Chinese peasant, I am only barely keeping my head above water. They continue to get in the way of less important things like work and continue to marvel at the ability of academics to ask stupid questions not once but time and time again. Some people never learn.

Chapter 1

Rural lives and rural studies

Introduction

Traditionally, agriculture lay at the very centre of rural studies. Most people in rural areas worked in agriculture and agricultural production, and the demands of agricultural production, for tools, services and processing, dominated the rural landscape. To talk of the rural was to talk about agriculture and all that it entailed, spanning the technologies of production through to the politics of land and the distinctive cultures and societies of the countryside. Agriculture, and therefore the countryside, became a place where society and science, culture and economy, and politics and production intersected and interweaved. The Dutch rural sociologist Timmer, in 1949, wrote that the 'countryside forms, as it were, a stage upon which, for the world, a very important play is performed; this play is called agriculture, and the head role is played by the farmer' (quoted in Ploeg 1993: 242). Thus for Timmer, and many other scholars of the rural, the countryside was all about agriculture, and in social terms agriculture was all about farmers.

But in the developed world agriculture no longer dominates rural areas, whether in terms of production or employment. Distinctive rural societies have apparently evaporated as rural areas have been infiltrated by new classes with new desires and agendas. As a result rural studies, as an area of academic enquiry, has had to contend with the decline of the centrepiece that informed, guided and justified its very existence. While some scholars have pronounced the death of rural studies – and rural geography – as a separate branch of academic endeavour, others have used the opportunity to reorientate rural scholarship. This has taken the study of the countryside in different, but not necessarily mutually exclusive, directions.

Some scholars, including Raynolds (1997), Raynolds et al. (1993) and McMichael (1997), have focused on the importance of global agro-food systems and concerned themselves with the manner in which rural areas and agriculture have been vertically and horizontally integrated into wider, global structures, partially erasing their distinctiveness and undermining their autonomy. Others, such as Hart (1997) and Goodman and Watts (1994), have argued for the continued exceptionalism of agriculture even in the context of global economic change, and stress the importance of locality. Using political economy perspectives derived from industrial geography, scholars like Marsden et al. (1993) have injected a concern for social change and class analysis in the new rural economy, while Cloke and Goodwin (1992), and others, have taken this in a slightly different direction to focus on the 'restructuring' of rural areas and the importance of consumption in the new rural spaces.

This reinvigoration or reorientation – not all scholars are altogether sanguine about the directions that have been taken (see Miller 1996)[1] – of rural studies in the developed world, which essentially dates from the mid-1960s, has only partially extended to work in the developing world. This is largely because rural areas of the developing world would appear not to have undergone any fundamental process of restructuring. Agriculture still dominates the country-side both in terms of production and employment. Furthermore, agriculture continues to be viewed as the core sector in much of the developing world. And finally, the idea that there is a distinctive culture and society in rural areas remains, in superficial terms at least, convincing.

This is not to say that developing world rural studies has stood still. Far from it. Miller's assertion that 'What is currently lacking in rural studies – and urgently required – are bounded cases, *monographs* . . .' (1996: 111) does not resonate with the developing world where the detailed, bounded village or community study has been a scholarly rite of passage for years. Indeed, in a number of significant respects rural work in the developing world has been in the vanguard, for example, in its concern for class and class differentiation, gender, moral economies and weapons of the weak, and the role and place of NGOs in rural change. So it is not a question of rural studies in the South being 'behind' (or 'in front') of work in the North. Rather, many of the concerns that have directed scholars working in the countries of the South have been distinct and different from those that have exercised the minds of scholars focusing on the North.

Constructing rural society, constructing rurality

Rural geography and rural studies in general are founded on the assumption that there is something special about the rural which makes it distinctive and therefore different from the urban. 'The concept of the rural', Mormont writes, 'evolved by distinguishing the rural and the agricultural, and by defining the rural in relation to the social and cultural context created by industrial develop-ment . . .' (Mormont 1990: 22). This view of the rural as different and distinctive is problematic in a number of respects.

To begin with, there are compelling reasons to question the belief, embed-ded in rural sociology, that industrial change has led to the emergence of two independent – or 'autonomous' – worlds, rural and urban (Mormont 1990: 21). As the discussion in Chapter 3 shows, historical processes of change in Southeast Asia have frequently been fluid and their outcomes ambiguous.

[1] Miller writes in his fierce critique of Cloke *et al.*'s (1994) *Writing the rural*: 'The only meaningful resonance here is with the painful and debilitating experience of having to read *Writing the rural*. The only real function of such a literary catastrophe can . . . be to act as a siren, a shrill warning to the innocent against the seductions of undisciplined reflexivity and derivative theorizing' (1996: 109). See also Cloke (1996) for a reply.

Moreover, the cultures that constitute the region have been embroiled in a continual historical process of formation and reformation,[2] with historians embracing a view of the past that stresses complexity and multiplicity, so any simple assumptions about the direction of change, and about spatial and social cleavages, have been called into question.

Second, historical research has contested the image of rurality that rural sociology has foisted on the countryside. In this view of things, it is the particular cultural and moral milieu that constitutes rural society which makes rural people distinctive – their concern for family and community, their 'moral' economy of sharing and communal support, their conservativeness, and their self-reliance and dislocation from the mainstream. So, while rural societies across the world are different and diverse, they nonetheless share certain common features that permit scholars to distinguish rural society from urban society. Significantly, the roots of difference in this schema are not socio-economic but socio-cultural. Moreover, this distinction between rural people/society and urban people/society is not simply an indulgence of scholars. There is an ingrained belief among rural (and urban) people that they are different. Mills, for example, quotes a textile factory worker from the countryside living in the Thai capital:

> **People in the city and people in the village aren't the same. City people, Bangkok people, you can't trust them, they only think of themselves. In the city people don't know each other. I've lived in this room for many months now and I still don't know the neighbors. In the village, I know everyone. We grow up together, we're all relatives and friends together. I know where they come from, their background. I can trust them. (Mills 1997: 48).**

In the light of this textile worker's comments, it could be argued that critics of the 'rural' and their attempts to challenge and erase its distinctiveness ignore the significance of the countryside as an idea. The rural may no longer be very distinctive from the urban in terms of economy and society, but (some) rural people believe it to be so. Facts are not enough. Just as the past is reinvented to serve the needs of the present, so the present is also given an ideological gloss. Nostalgia for the past informs the present, and vice versa.

A third problem, and one that arises directly from the 'two worlds' thesis, is the parallel position that these two worlds – and the people and societies therein – embody different, and often conflicting, interests. In the context of the developing world, this is best reflected in the voluminous literature on urban bias which seeks to characterise rural areas and rural people as in competition with urban areas and urban people (see Rigg (1997) and Chapter 9). The urban bias thesis homogenises and stereotypes rural society and sets it in opposition to urban society. This is then employed as an overarching explanation for the continuing poverty and underdevelopment of the countryside. Not only is the urban bias thesis reductionist, it is also static. The possibility that processes of rural (and urban) change might fundamentally challenge the roots of urban bias is not seriously entertained.

[2] See also Berry (1993) on Africa.

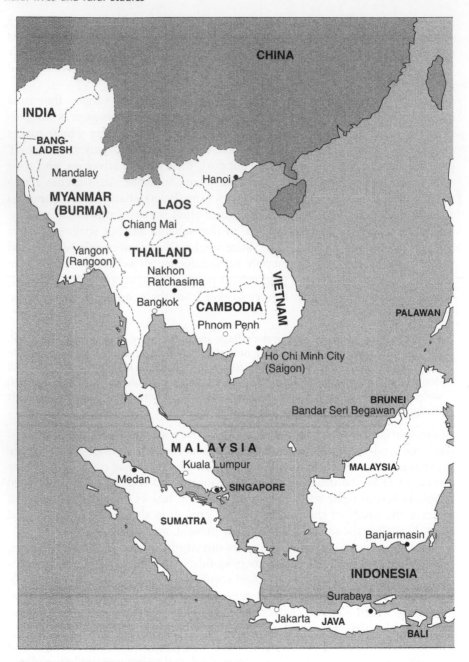

Figure 1.1   Southeast Asia

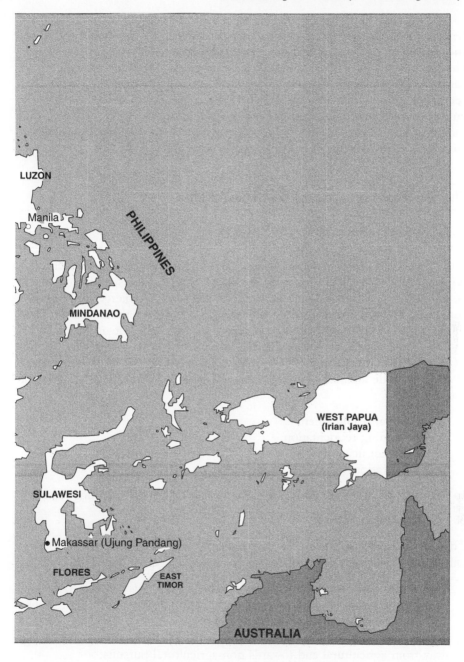

The question of change highlights a deeper shortcoming of rural studies: a tendency towards static interpretations of the countryside. Even should one accept the visions of tradition which have been purveyed as emblematic of rurality, change has rendered them into historical artefacts. The interpenetration of rural and urban reflected in the infiltration of new activities into the countryside, the far-reaching social changes that have been associated with the pervasive culture of modernity and the spreading availability of education, and the increased mobility of rural people embodied in the high levels of commuting and circulation, all point towards the need to reinterpret rural society and economy.

The de-agrarianisation of Southeast Asia

A large portion of this book is about the de-agrarianisation of Southeast Asia. While the focus is regional, the argument is constructed from the bottom up, using largely local-level studies, many taking a qualitative approach. The lessons to be learnt from these studies are diverse and the specific local-level contexts play a defining role in dictating the pattern that emerges in any given rural area. But while the specifics are varied and complex, the broad direction of change – the historical sweep – is clear and it is one of de-agrarianisation.

De-agrarianisation is a term that has come to be used to describe a series of parallel changes in rural areas of the developing world. In her discussion (1997a: 4) of de-agrarianisation in Africa, Bryceson highlights four facets of the process:

1. Occupational readjustment
2. Income-earning reorientation
3. Social re-identification
4. Spatial relocation

In the case of Southeast (and East) Asia we can add a fifth defining feature:

5. Spatial interpenetration

The forces propelling this process are numerous, as the discussion that follows will show. However they can be corralled under five broad headings:

1. **Economic** The changing balance of return to labour between farm and non-farm activities and the widening availability of non-farm opportunities to people living, or attached to, rural areas.
2. **Social and cultural** The shift in aspirations, especially among the young, away from agricultural and towards non-agricultural pursuits.
3. **Political** The prevailing neo-liberal culture and the emergence of the structural conditions, globally, nationally and locally, which promote spatial economic integration, industrialisation and, more particularly, rural industrialisation.
4. **Infrastructural** The markedly improved network of communications and the corresponding increase in mobility which has led to a transport revolution in rural areas.

5. **Environmental** The decline of agricultural potential associated with emerging environmental problems.

These forces are not discrete but closely interrelated. For example, the decline in the environmental integrity of rural areas is sometimes linked to the intrusion of industrial activities. This is associated with improvements in transport infrastructure and with the availability of labour with the wish and the skills (education) to become factory workers. This, in turn, needs to be viewed in the context of the national and global political economies which are creating the conditions where it is possible and attractive for firms to locate in rural areas.

The de-agrarianisation of Southeast Asia (and other regions of the developing world) cannot be equated with the restructuring thesis applied to the countries of the developed North. Counterurbanisation, for example – an important element of the debate (see Halfacree 1994) – is only weakly represented.[3] Nonetheless, in so far as all countries are participants in a process of global restructuring, there are clear points of contact between the North and the South.

Tensions

While much of the empirical content of this book focuses on the interrelated processes that constitute de-agrarianisation, at another level the discussion returns, time and again, to the emerging tensions in rural society, economy and space. However, the use of the term 'tension' here is not meant to have pejorative or negative and destructive associations – or not necessarily so. Indeed, many of the tensions can be regarded as a healthy contest between different activites, different social outlooks, and different positions.

In terms of agricultural production, there are evident tensions between the forces of globalisation and those of subsistence; between profit and sufficiency; between local needs and global demand. Widening the perspective, there are also tensions between the requirements of agricultural production and those of non-farm production; of environmental sustainability over the maximisation of return; of farm work over non-farm work. These tensions in the sphere of production both reflect and amplify tensions in rural society: between the family and the individual; between men and women; between the young and the not-so-young; between the landed and the landless; and between the rich and the not-so-rich. These social and economic tensions have, in turn, various spatial manifestations as industrial production infiltrates the countryside and farmers colonise urban areas.

These tensions, and the agrarian structures that underpin them, have divergent, locally unique outcomes. This raises another tension which is clear in the discussion that follows: the tension between global processes and local outcomes, between the recognition that agrarian change is non-linear and locally

[3] Although work on Extended Metropolitan Regions, discussed in later chapters, highlights the blurring of the divide.

unique, and the belief that there is at the same time a direction of change that can be identified and highlighted. The book never satisfactorily reconciles this final tension, perhaps because it is endemic. Local communities are not 'outside' global processes; they are part and parcel of such processes and engaged in their definition. With this in mind, the focus here is very much on the people and their histories and cultures, and their place and role in the global processes and structures that influence lives but do not govern outcomes.

Chapter 2

Agrarian questions for a new rural South

Introduction

After many years of theoretical neglect, the 'agrarian question' has latterly been the subject of considerable scholarly attention and debate. Perhaps the fullest treatment is Byres' authoritative *Capitalism from above and capitalism from below: an essay in comparative political economy* (1996). As he points out, Karl Kautsky's original work *Die Agrarfrage* ('The Agrarian Question') published in 1899 only received the imprimatur of a full English language translation nearly a century later, in 1988 (Byres 1996: 37, fn 5). Until then scholars wishing to work with an English language version were forced to turn to Banaji's (1976) 50-page summary of Kautsky's work.[1] In geography, Kautsky came into vogue during the 1980s, by way of rural sociology (Marsden *et al.* 1996: 363).

While the agrarian question may have become a central issue of debate, it has done so in various guises. Some authors have pronounced that the agrarian question, at least in the developed world, has been 'resolved'. Others have sought to 'redefine', 'reconstitute' or 'renew' the agrarian question, making it relevant to the issues of the day. But in redefining the question there are, again, different priorities at work. For Ploeg (1993), for example, the 'new' agrarian question is how to ensure that agricultural science does not forget rural sociology. For others the 'question' is one of redefinition in a context where global processes are, so it is said, dictating local outcomes (see Watts and Goodman 1997: 6). And yet other scholars – such as Bernstein (1996) – believe that the new possibilities for industrialisation in the South brought about by globalisation mean that the question is no longer pertinent.

The intention in this chapter is to set out the broad parameters of the agrarian question as it was classically formulated at the turn of the twentieth century. Rather more attention will be paid to the issue of the relevance of Kautsky's work to contemporary rural change. For, and as will become clear, the renewed interest in the agrarian question stems from its potential application, and its theoretical resonance (or lack thereof) with current debates. While a significant proportion of this attention has focused on rural restructuring in the developed world, scholars have also sought to illuminate its relevance to processes of change in the developing world (see for example, Pincus 1996 on Java).[2]

[1] A fact that Shanin and Alavi find 'problematic' (1990: fn 10, p. 260).
[2] There is a good case for arguing that the developed/developing world division should be rejected in any case.

Kaustky's concern with the political economy of agrarian change also provides a historical link with the so-called 'new political economy of agriculture'. This has sought to apply conceptual advances in the study of industrial change to farming. Of particular concern has been the question of the influence of global agro-food systems and the role of international capital in transforming agriculture and rural areas, both in the South and the North. This attempt to interpret agrarian change as an outcome of processes mediated in the global arena has, in turn, led some scholars to react by rekindling the debate over structure and agency and to argue that locality continues to have a determining role in dictating patterns and processes of change.

A third area of debate which will be addressed here – albeit briefly – is the issue of rural differentiation. Once again, this has its roots in work conducted mainly in the developed world, where authors have identified a fundamental shift in the relationship between rural and urban areas, and between agriculture and industry. In essence, rural areas, for so long regarded as marginal and distant from the main centres of social and economic change, now occupy a central position. Processes of rural industrialisation, counterurbanisation and the associated tight vertical and horizontal links that are being established between economic activities and populations have served to erode, so it is argued, long-established assumptions of difference.

The classical agrarian question

We should ask: is capital, and in what ways is capital, taking hold of agriculture, revolutionizing it, smashing the old forms of production and of poverty and establishing the new forms which must succeed? (Kautsky [1899] in Banaji 1976: 3).

Agricultural production has already been transformed into industrial production. . . . Economic life even in the open countryside, once trapped in eternally rigid routines, is now caught up in the constant revolution which is the hallmark of the capitalist mode of production (Kautsky [1899] in Goodman and Watts 1994: 1).

The agrarian question usually, and classically, concerns agriculture, capitalist accumulation and the transition to capitalism. More particularly, it embraces the introduction of capitalist relations into peasant agriculture, the associated transformation of agricultural production, and the role that agriculture – so transformed – plays in industrial development (Bernstein 1996: 29). This latter point is important for while the focus of the agrarian transition is on the countryside, it embraces, in the broadest sense, 'those changes in the countryside of a poor country necessary to the *overall* development of capitalism and the ultimate dominance of that mode of production . . .' (Byres 1991: 3–4 (emphasis in original)). One of the key outcomes of capitalist transition in the countryside is de-peasantisation, and the transformation of social classes under the influence of capitalist transition is a central component

Plate 1 A farmer in Ban Hua Rin in the province of Chiang Mai, Northern Thailand looks out over his soon-to-be-harvested rice. He rents a little less than 2 hectares of riceland from an absentee landlord. The rice here is a Japanese variety that a subsidiary of a brewery company has contracted the farmer to grow. It will be used to brew sake.

of the transition itself. Thus the agrarian question is both an economic and a political question.[3]

However, while the broad, historical picture may be clear enough, the paths of transition vary tremendously. The resolution of the agrarian question in every country is exceptional and, indeed, for each country there are many paths to resolution (see Byres 1991, White 1989). A second important point, which Byres (1995: 566) notes, is that the agrarian question is a dynamic one. And it is dynamic in two major senses. First because it is a question (really, questions) posed against a backcloth of change embracing society (peasant differentiation), class (landlord and peasant capitalism), technology (machinery, new crops and varieties of crops, chemical inputs), and the state (land reform, various other state orchestrated interventions). And second, because this backcloth necessitates that the question itself be recast as changing conditions dictate.

It has become common, drawing on the work of Lenin, to draw a distinction between 'capitalism from above' (reactionary) and 'capitalism from below'

---

[3] As Shanin and Alavi note (1990), in the opening chapters of *The agrarian question* Kautsky presumes that the forces of capital accumulation in the countryside will eliminate the peasantry. However, later in the book Kautsky finds himself attempting to explain why the peasantry continue to persist even in the face of capitalist development. So, while Kautsky continues to identify de-peasantisation as the inevitable outcome of capitalist penetration into peasant agriculture, there is an important change in the assumptions that underpin the argument.

(popular).[4] In the United States both forms, broadly, can be seen in historical operation. In the South, slavery was embraced to overcome severe labour shortages, but this in turn created new obstacles to accumulation, in particular an absence of incentives and a tendency to neglect technological advances. The American Civil War swept this system away but it was only with a new era of labour shortages during the Second World War that the agrarian transition could be successfully completed. In this instance labour shortages were met by mechanisation, not by slavery. Wage labour was displaced from agriculture and a class of capitalist farmers emerged. While this does not conform to the classic formulation of 'capitalism from above', it can be contrasted with the situation in the North and West of the United States. Here, the agrarian transition was much closer to Lenin's 'capitalism from below'. Family farms, private entrepreneurs, rising land prices and increasing market integration created conditions for intensive commodification of production. Hampered by a lack of labour, farmers turned to technological innovations, from barbed wire to machinery, to lower unit costs and raise productivity. A system, as Byres puts it, of 'advanced simple commodity production' emerged (Byres 1996: 371).[5]

The situation as it evolved in Prussia came closest to Lenin's 'capitalism from above' – a highly reactionary working out of the agrarian question in which feudal landlords largely dictated the course of events, preventing the emergence of a vital peasant economy based upon peasant differentiation. In the process, feudal landlords and tenants were transformed into an agrarian capitalist class and wage labour, with just a small number of richer peasants making the transition to becoming capitalist farmers (Akram-Lodhi 1998: 136, Byres 1991: 22–27).

Thus, while there may be one question, this question has been resolved in the developed world in different ways.[6] Not only are there Prussian and American paths to agrarian transition, but English, French, Japanese and South Korean/Taiwanese too. Nor is it likely that even this smorgasbord of models is either exhaustive or offers much of an insight – except at a very general level – into likely or actual agrarian transitions in the countries of the South (see the next section).

The classical agrarian question and the rural South

At the outset it should, perhaps, be noted that the discussion for the rest of this chapter is drawn from a diverse series of papers and books which often present highly nuanced interpretations of the processes of economic change

---

[4] A detailed re-examination of the agrarian transition in the United States and Prussia takes up the bulk of Byres' *Capitalism from above, and capitalism from below: an essay in comparative political economy* (1996).

[5] The image of American pioneer farmers occupying unpopulated land does not take account of the displacement and, in some cases, obliteration of Native Americans (American Indians) in the area.

[6] Or, it appears to have been resolved. Bernstein makes the point that in Byres' reading of events, agriculture in Prussia and the South of the United States did not contribute to capitalist transformation, and therefore they cannot be viewed as instances of successful agrarian transition. Accumulation in agriculture did not arise and capitalist transformation occurred *independent* of an agrarian transition (Bernstein 1996).

currently under way in the countries of the South and North. The intention is not to provide a summary of these arguments but rather to pick out, magpie-like, those aspects that have a bearing on the overall issue addressed here – namely, how far the countries of the developing world, as their agricultural sectors evolve, face a different set of contexts, questions and challenges from those that confronted the developed world.

As was noted in the introduction, in the industrialised world the agrarian question no longer occupies centre stage. For some scholars this is because the question has been successfully resolved; for others it is because agriculture is no longer a key activity. In much of the developing world, the reverse is the case: the transition to capitalism is an ongoing project and, notwithstanding far-reaching structural change in some countries, agriculture remains an import-ant economic sector employing the largest proportion of the workforce. Thus we have, on the one hand, a completed historical process in the developed North, while in the South the process is contemporary and continuing. In the North obstacles to accumulation were overcome and capitalist transformation and industrialisation ensued. In the South, significant obstacles remain. On the face of it, then, it would appear that the agrarian question remains highly pertinent in the developing world.

Byres makes the point (1991: 59–60), noted above, that the sheer diversity of historical paths to agrarian transition should warn us against simplistic models of change. 'Moreover', he writes, 'the history of successful transitions does not exhaust the *possibilities* in contemporary circumstances' (1991: 59 (emphasis in original)). This closely relates to a second issue, that of timing (Bernstein 1996: 42). With the exceptions of Taiwan, South Korea and Japan, the examples of so-styled successful transition completed this process before the end of the nineteenth century. The countries of the South had to wait until independence, between the late 1940s and 1960s, before they could even begin their journeys to transition. This is not a reason to ignore historical antecedents when it comes to interpreting current change, but it does warn against using history in a prescriptive fashion.

Most scholars would accept the 'continuing existence in the countryside of poor countries of substantive obstacles to an unleashing of the forces capable of generating economic development . . .' (Byres 1995: 569). The forces of accumulation have not been sufficient to bring economic development to rural areas and the agrarian question remains to be resolved. But there is a sig-nificant difference of opinion when it comes to suggesting that 'the agrarian transition not only has implications for the fate of the countryside, but has a decisive influence upon the pace, manner, limits, and very possibility of capitalist industrialisation' (Byres 1995: 569). In short, is it possible to achieve capitalist transformation without an agrarian transition? Where scholars stand on this question depends, in large measure, on their interpretation of the role of the global economy in national development. Those countries that have completed their agrarian transition were operating in a world system where national economies were dominant. Countries undergoing agrarian trans-ition today are functioning in a very different global context. Most clearly, the possibilities for foreign investment are such that industrialisation can occur independent of an agrarian transition and is likely to be 'mediated by the (differential) effects of the circuits of international capital and world markets . . .' (Bernstein 1996: 42–3). In this era of globalisation, the argument goes, countries are presented with the opportunity of industrialising without undergoing an agrarian transition, thus marking 'the end of the agrarian question *without its resolution*' (Bernstein 1996: 50 (emphasis in original), see also Marsden *et al.* 1990: 6–7).[7]

Successful agrarian transitions in Asia – models for the South?

There has emerged a tendency to regard agrarian transitions in Japan and, more latterly, in South Korea and Taiwan as indicative of the path(s) that other countries in Asia will, over time, follow. However, a closer examination of these three examples reveals special features that would seem to support the case for exceptionalism.

In Japan the landlord class, before the US-imposed reforms of 1945, were in the most part non-absentee landowners, resident in the same locality as their land was concentrated. They were also closely interested in issues of productivity – both features which, taken together, arguably make the Japanese experience unique. A third distinctive feature of the Japanese transition was the instrumental role played by the state. This extended from the state-orchestrated Meiji restoration of the late nineteenth century when classical feudalism was deconstructed, to the imposition of very high rates of taxation on the peasantry and the success that the state enjoyed in suppressing peasant disturbances (Byres 1991: 48–50). Finally, the American-imposed land reforms that followed the end of the Second World War obliterated landlordism and replaced it with a system founded on family-based owner-occupiers.

Just as any reading of Japan's agrarian transition must acknowledge certain unique (and central) features, the same is true of Taiwan and South Korea. Like Japan, the state – in this case the colonial state – had a defining influence. Japan colonised Taiwan and South Korea in 1895 and 1910 respectively and its role reflected Japan's own experiences at home, in particular in clarifying property rights and introducing a landlord–tenant structure. This permitted the thoroughgoing commercialisation of agriculture. Also as in Japan, the post-1945 period saw extensive land reform in both countries, once again propelled by external forces.

An area of continuing debate concerns the role of the post-1945 state in Taiwan and South Korea in directing the agrarian transition. For Amsden (1979) in Taiwan and Wade (1983) in South Korea, while agriculture was based in each case on a class of small-scale, family-based, owner-occupiers, the state played a defining role in agricultural change. Wade goes so far as to intimate

---

[7] Bernstein writes that 'the classic agrarian question is no longer of concern to capital on a world scale' (Bernstein 1996: 40), continuing in an endnote that the 'passing of the classic agrarian question, *from the viewpoint of global capital*, is reflected in the virtual disappearance of (bourgeois) land reform from the agenda of "agricultural development policy" since the 1950s and 1960s' (1996: 47 (emphasis in original)).

that, in his view, it makes as much sense to describe South Korean agriculture as 'one farm' as it does to characterise it as 'small-scale family farming' 'so tightly constrained is the area of farmers' free decision-making by the monopoly and monopsony powers of the state' (1983: 25). Amsden also notes the tight interpenetration of public and private interests under Taiwan's system of bureaucratic capitalism (1979: 362). So, and notwithstanding the numerical domination of the countryside by individualistic small farmers, the Taiwanese and South Korean experiences are closest to being state-directed agrarian transitions from above (Byres 1991: 58).

Even this cursory examination of the historical experiences of Japan and Taiwan/South Korea reveals the difficulties connected with using them as potential models for agrarian transformation in the other countries of Asia, or the developing world as a whole. The tyranny of the unique in these three cases would, yet again, warn against anything more than the loosest application.

New agrarian questions

Not only does the agrarian question have to be contextually defined according to the specific conditions – historical, social and economic – in operation, but there is also considerable speculation as to whether, in its classical formulation, the question is particularly illuminating in the context of the contemporary rural South. This is reflected in a series of papers which have sought to 'revisit' (McMichael 1997), 'recast' (Roberts 1996), 'renew' and 'repose' (Watts and Goodman 1997: 6) and 'go beyond' (Roseberry 1993) the agrarian question. The starting point for these papers is the suggestion that the characteristics of rural areas, rural production and rural people today, and in particular their intersection with the global economy, demand a rethinking of the assumptions that lie behind the original question. One of the criticisms levelled at Byres' work on the agrarian question is that it is 'peculiarly narrow' in ignoring the effects of globalisation (Watts and Goodman 1997: 6).

In geography, there has been a concerted attempt, as it were, to 'take the farmyard out of rural studies', resulting in the emergence since the 1980s of the 'new political economy of agriculture'. Hitherto, concern with the minutiae of production – fertilisers, new crop varieties, labour inputs and the like – has tended to sideline agricultural and rural geography at a time when economic geography more widely, and especially industrial geography, was drawing inspiration from political economy and, in the process, flourishing. This project to reinvigorate rural and agricultural geography is reflected, for example, in the tendency to view production as embedded in webs of social, economic, political and cultural processes and practices (Marsden *et al.* 1996). It also explains the concerted effort to apply Fordist and post-Fordist perspectives, and especially regulation theory to rural/agricultural geography (Cloke and Goodwin 1992 and see Page 1996). In these terms, the agrarian question too has been reformulated from its classical origins to focus on the transition of agriculture to Fordism and post-Fordism.

Importing, applying and drawing inspiration from ideas derived from the new economic geography, and in particular industrial geography, the new political economy of agriculture school has thus sought to reposition the study of agriculture and agrarian systems. In its most radical form, this perspective embraces the view that global restructuring has largely erased the distinction between agricultural and industrial paths of change and that, for example, Fordist models of industrial organisation are applicable and appropriate to understanding agriculture (see Goodman and Watts 1994: 4).

Another related trend has been to bring critical perspectives to bear on rural geography. For one of the outcomes of the concentration on farming systems – a productivist approach with empirical, rational and scientific overtones – was a tendency to overlook the issues of people, class and gender in making the social spaces of the countryside (see Phillips 1998: 41–2, Marsden *et al.* 1993). This criticism, it can be argued, has rather less relevance for studies conducted in the developing world, where geographers and anthropologists have maintained close intellectual links and where issues of difference have occupied a central position in studies of rural change. Further, these studies have not just investigated issues of difference for their own sake, but also suggested mechanisms by which such differences are maintained, accentuated and resisted. Perhaps the best known of such work is that which comes under the broad rubric of subaltern studies, including work on everyday forms of peasant resistance.

Agro-industrialisation and the persistence of the family farm

Classically, the agrarian question seeks to explain the transition to capitalism and the transformation or replacement of peasant farms by agro-industrial enterprises. But the surprising persistence and resilience of family-based production units, even in the developed world, has caused scholars to re-examine the assumption that there would be an inexorable advance of capitalist relations in rural areas.

What explains the persistence of the family farm and non-corporate activity – simple commodity production – in agriculture? (see Marsden *et al.* 1996: 361–2). Essentially, it has been argued that, for various reasons, farming is not conducive to capitalist transformation. Family farms are at least as productive and efficient as large-scale agro-industrial enterprises, and sometimes more so. The nature of the farming cycle, where production time exceeds labour time;[8] the challenges presented by variations (spatial and temporal) in environmental conditions; the dependence of agriculture on a finite resource (land); and the social and economic flexibility encapsulated in the family farm give it distinct advantages over more 'industrial' alternatives (Page 1996: 380, Roberts 1996: 399–400, Baxter and Mann 1992). Furthermore, the carrot of inheritance has been said to permit landowners to underpay family labour while farmers do

[8] With the result that 'farming labor cannot be applied constantly to production, thereby limiting surplus extraction . . .' (Page 1996: 380).

not have to contend with all the difficulties of supervising hired labour. The return on capital is not sufficient to entice large enterprises into agricultural production, leaving the family farm with the economic space to operate and – in most cases – to flourish.

The social and economic dynamism inherent in the family farm, and in particular the tendency for farms to be periodically regenerated, is also seen as an important component of agriculture's distinctiveness. Moran *et al.* note the necessity to acknowledge the links between the farm firm and the farm family (1993: 32–33). The farm family life cycle has important implications for the operation of the farm firm. This applies, for example, to labour availability in agriculture (young families are more likely to be in labour deficit) and patterns of consumption and expenditure (young families may have fewer avenues for income generation and face greater expenditure demands). The intergenerational transfer of the farm also brings a unique quality to agriculture. Sometimes this is characterised as the passing-on of the farm from one generation to the next, where the farm as a system of production remains largely unchanged. In other instances scholars stress that the farm is not so much 'passed on', as 'reconstituted' (Roberts 1996: 407–8).

But, while the family farm may continue to dominate land-based production, agro-industrialisation has encroached on the farm from below and above. Unable to transform farming for the reasons outlined above, capital has, instead, sought to capture surplus value by the parallel processes of 'appropriation' from below and 'substitution' from above (Goodman *et al.* 1987). Inputs into the farm system – agro-chemicals, new seeds, animal feeds, mechanisation and the like – have been 'appropriated' by manufacturers who produce these inputs under industrial conditions for supply to farms. At the same time, purchasers of farm outputs increasingly demand only the basic commodity, with all further processing undertaken off-farm, and again under industrial conditions. Historically, this process of substitutionism can be seen in the advance of flour milling, canning and refrigeration, and the industrial dairy industry and, most starkly, in the substitution of agricultural commodities for artificial products like margarine and artificial fibres (Goodman *et al.* 1987). As this encroachment proceeds 'the extent of industrial activities linked to agriculture expands greatly, and industrialists and merchants increase their effective control (directly or indirectly) over on-farm labor' (Page 1996: 381–2). It is by this means, it has been suggested, that capital extracts surplus value from agriculture. The family farm, in a sense, becomes a 'shell' while the truly profitable aspects of farming are appropriated by industry. It survives, but only as a 'nexus that manages and takes risks for labor and natural processes not yet appropriated as industrial inputs' (Roberts 1996: 400).

Another perspective, which has not been widely developed in the conceptual literature in agricultural geography but which is particularly pertinent to Southeast Asia, is the issue of the collapse of the 'farm family' as an internally identifiable and coherent unit of social and economic analysis. As will be demonstrated in later chapters, it is the involvement of farm families in non-farm industrial activities that permits small-scale family farms to persist.

In other words, farms are drawn into industrial processes not just through appropriation and substitution but indirectly, through the income that farm households generate from non-farm activities. It is this combination of farm and non-farm that explains the surprising resilience of the family farm – and in particular, the sub-livelihood family farm.

There is a conceptual and practical tension between the (so-styled) universalising effects of globalisation and the uniqueness associated with locality. It is in agriculture that this is most profound. The social constructions that underpin family farming, the variabilities associated with land and environment, and the particular policy contexts within which farms are managed all tend towards specificity and exceptionalism. This theme is a recurrent one in the literature on the political economy of agro-food systems.

Box 2.1   Contract farming and the global agro-food system

Contract farming, largely absent from the literature as a topic for discussion until the 1980s, represents a fine example of how the influence of the global agro-food system impinges on people and localities. Such restructuring, it has been argued, 'can only be fully comprehended as a global phenomenon' (Watts 1994b: 228). Watts notes the need to differentiate contract farming from simple marketing or labour contracts. 'Specifically, contract farming entails relations between growers and private or state enterprises that substitute for open market exchanges by linking nominally independent family farmers of widely variant assets with a central processing, export, or purchasing unit that regulates in advance price, production practices, product quality, and credit' (Watts 1994a: 26–27). But contractors do not purchase labour power; rather they buy the commodities which labour produces (Baxter and Mann 1992: 240). And in this sense there is an important distinction to be drawn with piece-rate work and other wage labour arrangements.

But while the family farm may persist, the farmer's autonomy is circumscribed. The buyer, having contracted to purchase a certain amount of output in advance at a certain price (or a price calculated on an agreed basis), has also bought a modicum of power over the grower. This includes decisions over production matters (inputs, crop management, labour use) and also over the quality of the final output. Given the relative power of the grower versus the contractor, it has become usual to see the relationship as highly unequal. As Watts writes, contract farming represents 'quite distinctive, if locally varied, social relations of production in which independent commodity producers [farmers/growers] are subordinated to "management" through a distinctive labor process' (1994a: 28). In extreme circumstances, farmers can be characterised as little more than piece workers.

Contract farming is a major means by which international capital has penetrated agriculture in the developing world. Multinationals may dictate the crop to be grown, the technology to be used, the price of the final product, and the basis on which that product will be purchased. Sometimes the same companies will provide the inputs, the capital (loans) and the knowledge and advice necessary for production. In such cases it is pertinent to question whether family farms, at least in so far as they are emblematic of independence, really continue to exist in the substantive sense of the word.

▶

Box 2.1 (continued)

It is clear that the historical and social conditions under which contract farming has arisen in the developing and developed worlds vary considerably (see the examples from sub-Saharan Africa in Little and Watts 1994). Nonetheless, these locally specific forms of contract farming have their origins, arguably, in the same set of global conditions and processes. 'New diets, a transnational integration of consumption circuits, and the enhanced profitability of agro-food investments have all encouraged new agricultures geared to international and, in some cases, affluent domestic markets' (Watts 1994a: 54). It can be suggested that contract farming represents an example of post-Fordism or flexible accumulation in agriculture. This is attractive because of the apparent links between industrial restructuring and contract farming. In particular, the disintegration of the production process to numerous sites; the flexibility inherent in such a diffuse system; the cost-cutting that can be achieved by dispersal of activity but maintenance of control (what has been termed the 'balkanisation' of the labour force (see Scott 1984);[1] and, more broadly, the intrusion of the global economy into farmers' lives. But, and this has been discussed in greater detail in the main body of this chapter, in emphasising these similarities there is sometimes a tendency to overlook the special attributes of agriculture.

[1] Scott's study of the women's dressmaking industry in the Greater Los Angeles region reveals a further balkanisation from sub-contractors to homeworkers. This is also driven by the need to cut costs and maintain wage discipline. Such fragmentation maintains competition between subcontractors or homeworkers, pushing down wages while preventing the emergence of worker solidarity (1984: 20–21).

Farming, the nation state and globalisation – a new dependency?

While it might be accurate to say that much of the developing world has yet to undergo an agrarian transition, it does not necessarily follow from this that the agrarian question (at least as classically framed) remains pertinent. Primarily this is because globalisation 'marks a decisive theoretical and practical break: old issues have disappeared into history, and new paradigms are required' (Akram-Lodhi 1998: 135). McMichael (1997) embeds his reworking of the agrarian question within a world-historical process and this, in turn, within the context of the current debate over post-developmentalism. For him, the classical agrarian question is dead because rural classes and the rural sector play such a minor part – globally – in national agendas. He suggests that the agrarian question at the end of the millennium is about how local people in rural areas resist the hegemony of the global.

This concern with the role of global processes in dictating local outcomes led, in the 1980s, to the emergence of the 'new internationalisation of agriculture school' (see Raynolds *et al.* 1993) or, sometimes, a 'new international division of labour in agriculture'. The latter was exemplified in the division between producers in the South, exporting labour-intensive luxury crops like flowers, exotic vegetables and shrimps, and producers in the North cultivating capital-intensive crops like basic food grains (McMichael and Myhre 1991: 93, Friedmann 1993). There are two, linked, components to this body of scholarship which sometimes comes under the rubric the 'transnationalisation of the food system'.

First, there is the question of the extent to which the internationalisation of agriculture has undermined the power and autonomy of the state. There has been a suggestion, for instance, that the outcome of global capital accumulation has been to transform national states into transnational states with the result that governments can no longer coherently determine policy (McMichael and Myhre 1991 and see Bernstein 1996: 45–6) and where farmers have been displaced from the national to the international stage (Kearney 1996). And second, there is the issue of how far global agro-food systems have affected farming and those engaged in farming, particularly in the South. Scholars have written of a need to 'recontextualise' and 'reposition' agriculture in the light of these changes, and in particular to view production as just one element in the global food system (Munton 1992: 27).

Scholars like Barkin (1990) and Raynolds *et al.* (1993) have argued that the internationalisation of Mexican agriculture has undermined the ability of the state strategically to determine the progress of agriculture and, by extension, the lives of those people involved in agricultural production. While during the early stage of internationalisation from the mid-1960s through to the debt crisis of 1982–84, the Mexican government did have a direct and influential role in the agricultural economy, since then this has diminished. 'In summary', Raynolds *et al.* write, 'there has been a movement from a form of internationalization that maintained a nationalist component to a form of internationalization that is much more oriented towards global markets.' They continue: 'The Mexican state is ceding its ability to at least partially regulate Mexican agriculture to powerful trading partners backed by transnational finance capital' (1993: 1110). Kearney (1996: 128) extends this vision to the general level when he proposes that agro-industrialisation in the developing world has effectively displaced farmers from the national context where the state holds centre stage, and replaced them within the global economic theatre where transnational corporations are the major actors (see also Friedmann 1993, who argues much the same).[9]

But while Mexico may be a good example of internationalisation diminishing the power of the state and reducing its ability decisively to intervene in agriculture, basing a generalised statement on the Mexican experience is not warranted. Drawing on the examples of Puerto Rico and the Dominican Republic for comparison, Raynolds *et al.* (1993 and see Raynolds 1997) demonstrate that Mexico may well be exceptional in showing these trends, and that extrapolating from them is a precarious exercise.[10] Different histories, political structures, and socioeconomic configurations produce heterogeneous forms of internationalisation.

---

[9] While Kearney's book is a general statement on rural society his field experiences, significantly, are drawn from Mexico and Poland – and largely, apparently, from the former.

[10] McMichael and Myhre, significantly, in arguing their universalist position state that 'Mexico is an exemplary (if not exceptional) case' (1991: 93).

All that can be said, at a general level, Raynolds *et al.* (1993: 1116–1118) conclude from their geographically limited comparative study, is that:

- there is a general crisis of food security;
- export agriculture has significantly expanded;
- the global context (broadly defined) configures agricultural policies;
- agriculture is being restructured to enhance international competitiveness.

Sanderson (1986), also working on Mexico, sees the internationalisation of agriculture in the South as much more insidious than merely reducing the autonomy of the nation state. Farmers are integrated into global systems, agriculture is homogenised as agribusiness dictates production decisions, and the tastes and diets of the world are moulded by those of the West.[11] For Sanderson, Mexico's agricultural crisis is a crisis of employment, nutrition, distribution and legitimation, and in all these senses is a product of international integration (1986: 274). In this respect, Sanderson echoes Barkin, who states that the integration of the country into the world economy has 'wrenched' people from their local communities into the national polity where they are 'increasingly subjugated to the designs of an international market' (1990: 3). It is not just that governments lose power; peasants, crucially, find their livelihoods under threat as they are sandwiched between downstream suppliers (providing, for example, patented genetic material, chemical inputs and advice) and upstream buyers (dictating size, appearance and presentation of the product).[12] Referring to the trade in exotic fruits and vegetables in general, Friedmann writes of the 'paradise of eternal strawberries and ornamental plants for rich consumers' depending on an 'underworld of social disruption and ecological irresponsibility' (1993: 54). In the final sentence of his book, Sanderson returns to Kautsky, if not by name then at least by implication when he writes: 'In the language of stabilization and "fiscal responsibility", the agricultural policy elite of the new administration has set the rural poor adrift and relegated the agrarian question to a back burner, where it will surely simmer' (1986: 284).[13]

In toto, proponents of the new internationalisation of agriculture school argue that governments in the South find their options for influencing national development increasingly circumscribed; that national self-sufficiency in basic food stuffs is compromised; that farmers are tied into unequal and dependent relationships with agribusiness; and that the destitution of vulnerable groups is the local outcome. For much of the period since the early 1980s the 'globalisation' paradigm dominated interpretations of agro-food systems (Ward and Almås 1997: 616–17).

---

[11] See also McMichael and Myhre (1991) on this.

[12] This essentially takes the work of Goodman *et al.* (1987) regarding appropriationism and substitutionism and applies it to the global context.

[13] Barkin's book proposes that a way out of Mexico's economic crisis (remembering that the book was completed in 1989 and published in 1990) is to embrace a new economic strategy – the 'war economy'. Essentially, Barkin suggests that domestic food production should be stimulated to replace imports. This would raise welfare, increase purchasing power and therefore also feed into industrial growth, supporting the expansion of labour-intensive manufacturing already underway.

Plate 2   A still-born housing estate in Sanpatong, Northern Thailand. This estate was conceived during Thailand's economic boom, and expired in the country's economic crisis. Formerly riceland, the land is now colonised by grass and shrubs and is only good for grazing cattle and drying rice.

Contesting the global

> . . . the label 'globalisation', with its allusions to outsourced international pro-
> duction and intra-firm integration, has become common currency in agro-food
> studies. Such debased usage often reflects an uncritical extension of concepts
> from industrial economics and economic geography (Watts and Goodman
> 1997: 15).

Not all scholars were – or are – seduced by the popular tendency to reduce all economic activity to just another slice of global activity and recently there has been something of a backlash against the perceived reductionism inherent in the globalisation paradigm. The dependency perspective that insinuates itself into much of the discussion, for example, where transnational corporations dictate terms and global culture dictates tastes, is contested. Goodman and Watts view these latter day *dependista* perspectives as suffering from many of the inadequacies of their original namesakes. 'A standard picture of North–South agrarian restructuring produces a one-dimensionality, gravely handicapped by their exclusive focus on Third World exploitation' (1994: 25). The reality of politics, national and international, is that it is complex and shifting. Countries that have not successfully industrialised cannot be lumped together into a single 'dependent' category with the expectation that the category will be internally coherent – what Bernstein (1996: 51) calls the 'logic of residual categorisation'.

Casting modern day agro-food systems as creating a new dependency of the South on the North caricatures the relationship between countries and the role of transnationals in the system. This is something that Dicken (1994: 122) acknowledges when he writes that while there may have been a relative shift in power away from states and towards transnationals, the degree of complexity and the level of contingency makes it difficult to say very much more than this. What does seem to be clear is that local institutions are relatively powerless in their relationships with transnationals. It is here that the asymmetries lie, not between transnationals and nation states (Dicken 1994: 123). As Marsden *et al.* write:

> **Globalization is treated as a contested process in which the disorganizing impact of transnational agro-food actors conditions rather than determines the actions of local producers and consumers. 'Analytical' space for social agency and local diversity is maintained . . . (1996: 367).**

The same charge of reductionism can be levelled at the tendency to argue that 'domestic social and political relations are increasingly shaped by global capital circuits' and, moreover, to suggest that there is a 'universality' in this transnationalisation process (McMichael and Myhre 1991: 84). While there are clearly aspects of global change which serve to undermine the particularities of space, the manner in which global processes impact on local spaces produces, at the same time, further particularities. In other words, we cannot understand change in rural areas of the South simply as an outcome of globally-determined processes. Locality remains important because outcomes under global change are predicated on local specificities, including those of a historical nature. This, of course, resonates with the wide-ranging debate over structure and agency in economic change (see below).

Critics of the universalising tendency inherent in the application of Fordist/post-Fordist studies of industrial change to agricultural transitions (while acccepting that conceptual advances in industrial geography do have some bearing on the understanding of agricultural change), focus on two key deficiencies. The first has as much to do with their application to industrial, as to agricultural change – namely, the reductionist and generalised interpretation of historical change, and in particular a tendency to collapse all industrial change into a Fordist/post-Fordist dualism (Page 1996: 379–380, Sayer 1989, Hart 1997).[14] As the discussion of Kautsky and the agrarian question above emphasises, detailed historical analyses make clear the point that there are many paths of agrarian transition (see Byres 1995 and 1996). The second objection is the assumption that agriculture and industry are likely to share a common future in terms of the structures and processes of change. 'The parallels between agriculture and industry', Goodman and Watts write, 'are radically over-drawn, forcing the analysis [of agriculture] into a theoretical straitjacket that leaves

---

[14] '. . . the literature on postfordism is confused in its arguments, long on speculation and hype, and based on selected example whose limited sectoral, spatial and temporal range is rarely acknowledged' (Sayer 1989: 666).

little room for diversity and differentiation within and between agrarian transitions' (1994: 5, see also Goodman 1997).[15]

This begs the question: why should agriculture be different? There is a persuasive case that the special characteristics of agricultural production and society demand a more cautious application of theoretical perspectives derived from industry (Goodman and Watts 1994: 37). There are, in other words, features of agricultural production that make agriculture a special case. Agro-industrialisation may be advancing globally, but this is not sufficient reason to make agriculture just another arm of industry. Agriculture remains place-specific, moulded by the particularities of history and environment, and the specificities of national policy regimes and agricultural commodity chains (see Marsden *et al.* 1993).

So, interpretations of rural/agricultural change that stress the global and embrace conceptual models created to explain changes in industrial geography and activity are inadequate in at least two senses: first, because agrarian transitions themselves are heterogeneous and contingent on unique configurations of history, society and economy (see Byres 1995 and 1996), and second, because agrarian and industrial transitions may be more different than they are alike. For Byres (1996), while agrarian transitions vary, the agrarian question, at least in the developed world, has been broadly resolved. Watts and Goodman disagree with this interpretation and instead argue that the agrarian question is 'alive and well', but living in a new global era (1997: 16).

Structure and agency in agriculture

The debate over the independence of family farms under conditions of contract farming (see Box 2.1 on page 18) raises the rather wider question of structure and agency in agriculture. A major source of dissatisfaction in geography with political economy approaches to agriculture and rural studies is the relegation of the individual and the household to the status of pawns on a structuralist chessboard. Prescriptive, unilinear and deterministic accounts of agrarian change 'failed to illuminate the diversity and adaptability of farming practices encountered in contemporary empirical studies' (Marsden *et al.* 1996: 363, see also Marsden *et al.* 1993, Hoggart 1992). While this may, in itself, be an oversimplification of the work of political economists like Byres and Bernstein who have emphasised the diversity of paths to agrarian transition (see above), there is certainly a sense in which human agency has been underplayed. It has been argued that the new political economy of agriculture, reflected for example in the work on global agro-food systems, shoehorns (self-evidently?) highly diverse and heterogeneous national food systems into a single global stucture (Goodman and Watts 1994: 20–21). There are, no doubt, transnational corporations operating

[15] Furthermore, it is not just the case that the particularities of agriculture, and different agricultural places bestow special characteristics on agricultural (as opposed to industrial) transitions. The power of the consumer to demand, for example, organic produce or humanely raised livestock also militates against Fordist production techniques.

Plate 3 The internet comes to Sanpatong, Northern Thailand.

within a global system, but this does not make it a necessity to argue that national producers have become divorced from the influence of the nation state. Global food systems and national uniqueness, not to mention human agency, are not mutually exclusive. Indeed, the intersection of the global and the local can lead to greater local distinctiveness, rather than less (see Amin and Thrift 1994).

This renewed interest in human agency is reflected in work on technological change in agriculture which has emphasised the diverse ways in which farmers and suppliers negotiate between themselves, resulting in diverse and highly localised outcomes (Marsden *et al.* 1996: 364). Ward, for example, objects to technological determinism where the progress of technology is seen as inevitable and independent and autonomous from society (Ward 1995: 19). Thus the questions to be asked are framed in terms of 'How can we best manage technology change?' rather than 'How can technology be best adapted to suit the needs and demands of society?' In his study of pollution from agricultural pesticides, Ward concludes: 'Technological change is socially shaped, as is the very nature of the environmental problem itself' (1995: 32). The actors in his study are farmers, advisors and the government.

Rural differentiation: putting it back together

Scholars of rural change in the developed world – and particularly in Britain – have emphasised the extent to which the 'rural', as an identifiable and coherent social and economic space, is breaking down (see Ilbery 1998b). The new rural space is characterised by multiplicity and competition between people with divergent interests. This takes two forms. On the one hand rural society, it is argued, has diverged and there are new patterns of alliance and competition between producers and consumers of rural space. Farmers (producers) have seen their relative political power diminishing during an era of food security and agricultural decline, just as consumers have seen theirs in the ascendancy.[16] Most rural dwellers have no direct links with agriculture. For them, rural areas are spaces for recreation, housing and conservation. Farming may have a residual hold on these people, but as a romantic construct or category, not as a production system (see Munton 1992: 44 and the chapters in Marsden *et al.* 1990). The second reason why the 'rural', as a category, is becoming increasingly problematic is because of the interpenetration of urban activities (i.e. industry or non-farm activities) into rural spaces (for a fuller discussion of this, see Chapter 8) and the delocalisation of work (Mormont 1990: 30–1 and Cloke 1998). As will be discussed in more detail in Chapter 4, the mobility of rural people takes them to work beyond their immediate place of residence, whether on a daily basis (commuting) or for longer periods (circular/seasonal movements). The reverse is also the case, although it is rather more prevalent in the developed world: urban residents consume the countryside for recreation. The result is that rural livelihoods increasingly depend on activities located beyond the immediate rural space or on non-farm (industrial) activities that have penetrated the countryside.

Consider this view on rural change in the developed world: '. . . the central organizing principles established in postwar times . . . have been largely overtaken by the tide of a rural (and urban) restructuring process which has been both economically and socially driven' (Marsden 1998: 15). This change is explained in terms of two principal shifts. A 'horizontal' shift of people and economic activity from urban to rural areas, leading to a process of counterurbanisation – a process which, as Marsden notes, has 'redefined the relationships between town and country . . .' (1998: 15). And a vertical shift of influence towards upstream and downstream producers and processors, and towards central agencies, whether national or international (e.g. the EU). What is perhaps most important in these changes is the *centrality* of rural space. No longer are rural areas in the developed world marginal places dominated by a marginal sector and inhabited by people engaged in marginal activities. But this change is not because agriculture has suddenly become newly important but because rural areas have been infiltrated by new people and activities that have more political muscle and economic weight.

From afar it can appear that rural areas of the developing world remain largely coherent and undifferentiated – that farmers, agriculture, the countryside and rural people are categories that share a common basis, and by extension common 'interests'. In other words, that we can assume that the countryside is dominated

---

[16] Munton writes of the tendency of social scientists 'to disregard developments within the [farming] industry, to treat them as intrinsically conservative or even vestigial' (1992: 26).

by agriculture, that rural people are largely farmers, and that farmers are primarily agriculturalists. This sequence of associations permits us, in turn, to be relatively cavalier and generalised in our descriptions and our interpretations of change. However, just as scholars working on rural areas in the developed world stress uneven development and heightened levels of differentiation (see the papers in Ilbery 1998a), so this is also true of the developing world and, in this instance, Southeast Asia. For Southeast Asia, it is not just a case of contemporary processes of change creating and accentuating difference. There has also been a re-evaluation of the past. Increasingly, historians are stressing the degree to which rural areas even in pre-modern Southeast Asia were characterised by difference.

In a provocative paper published in 1990 Hoggart suggested that the 'broad category "rural" is obfuscatory' and no matter how it is defined 'does not provide an appropriate abstraction' (Hoggart 1990: 245–6). His rejection of 'rural' in this way is based on a combination of diversity and difference. First, the evident diversity of rural areas makes it difficult to treat them as a category assuming some level of internal coherence. And second, the processes which underpin change in rural areas are, he suggests, not distinctive from those that are present in urban areas. Thus rural studies 'groups together places in which social processes of a very different kind are in operation'. Instead of choosing places to undertake research we should instead, he suggests, 'focus on particular social conditions (agency or structure) and evaluate how these unfold in particular settings' (1990: 254). While Hoggart's paper was important in highlighting significant changes in how rural areas are populated and structured, there have been strong relational links between city and country for many years. Therefore, the characterisation of 'the rural' and 'the urban' has always been more a convenient conceptual shorthand rather than an accurate reflection of reality.

In the context of rural areas of the developing world – and this is particularly true of Southeast Asia where the tendrils of globalisation have reached further into the countryside and deeper into the lives and livelihoods of the region's inhabitants – the agrarian question is not dead but renewed. Globalisation has not erased the importance of rural areas, nor has it erased the importance of locality. Processes of agrarian change are embedded, at the local level, in political, social and cultural relationships, as well as in environmental contexts through and upon which the forces of globalisation are reworked in unique and place-specific ways. There are those scholars who are interested not in globalisation as a process of erasure but as one which 'opens up "analytical" space for social agency and local diversity' (Whatmore 1994: 54, also see Amin and Thrift 1994, and Goodman and Watts 1994). Thus patterns of change in rural Southeast Asia are, at the same time, both similar to and different from those operating elsewhere. They are alike in the sense that the penetration of global capital (for example) into rural lives is a universal phenomenon; different in the sense that this process of penetration, and its effects, are unique.

Chapter 3

Delineating the rural past

Introduction

An understanding of present conditions in rural areas of Southeast Asia is both a product and a reflection of the past. History, in other words, has a bearing on contemporary change both substantively and conceptually. Substantively to the extent that the present is a product of a succession of historical processes, and conceptually in the sense that how we understand and interpret contemporary issues is a product of how the past is constructed. As Brown has noted in his history of economic change in Southeast Asia since 1830, 'our view of the pre-colonial rural order profoundly informs our understanding of the economic and social change which took place . . . during the colonial period' (1997: 10). It also, in turn, influences and informs our understanding of change since the colonial period. As the last chapter emphasised, agrarian transitions are critically contingent on unique configurations in which history plays a central part.

These twin issues of 'history' and 'constructed history' clearly have application worldwide. However, they are particularly pertinent to rural Southeast Asia because there are few areas of the world where interpretations of the rural past have been more contested. To quote Brown once more: 'Historians have long held sharply divergent views over the organization of production and the nature of socio-economic relations [in rural settlements] in pre-colonial South-East Asia' (1997: 8). Furthermore, these contesting interpretations have been fuelled, to a significant extent, by a desire to justify, explain or account for contemporary interventions. Or to put it another way, these apparently arcane debates over the past have been utilised – at a practical and applied level – to draw attention to the inequities of the present, to promote particular policies and programmes, and to justify certain ideological positions.[1] In his account of social and economic change in the Malaysian village of Sedaka, Scott writes:

> Having lived through this history, every villager is entitled, indeed required, to become something of a historian – a historian with an axe to grind. The point of such histories is not to produce a balanced or neutral assessment of the decade

[1] An example is Apichai Puntasen's (1996) paper on self-reliance in Thailand which draws on Chatthip Nartsupha's retrospective construction of the nature of the village in Thailand (see page 29). Apichai maps out the deleterious effects of modernisation on rural people and agricultural systems and on this basis calls for a self-reliant system based on local wisdom, integrated farming and the critical role of NGOs. Kitahara (1996) does much the same.

but rather to advance a claim, to levy praise and blame, and to justify or condemn the existing state of affairs (Scott 1985: 178).[2]

It is for these reasons that it is considered appropriate to allot a chapter in a book on contemporary change, to questions of historical change.

The nature of rural life and livelihood

What were rural settlements like before the modern period? How were they organised? What levels and types of interaction did they have with the outside world? How far were they subsistence-oriented? To what degree did the state intrude into village affairs? What level of commercialisation was evident? Were they egalitarian? Did concerns for the community over-ride those of the household and individual? Were they 'timeless', or did they respond to wider historical changes? These are some of the key questions that have divided scholars of the region – historians, geographers, economists, anthropologists, political economists, rural sociologists and developmentalists alike.

The village is often presented as the pre-historic building block of Southeast Asian societies: before modernity, before commercialisation, before cash crops and modern technology, before cities and trade, there was the village. This sense of villages – and the people who inhabit them – being disengaged, indeed immune from historical processes is evident in the opening pages of Dasgupta's book, where he writes of their having 'survived hundreds of years of wars, making and breaking up of empires, famines, floods and other natural disasters as the principal social and administrative unit' of the Third World (Dasgupta 1978: 1). Chatthip Nartsupha, one of Thailand's most influential economic historians, likewise writes of the village in Thailand as 'one of the most ancient institutions in Thai society', apparently insulated from the machinations and insinuations of the state, a 'self-sustaining and relatively autonomous unit' that 'existed in tranquillity throughout the long period of time which saw external changes . . . numerous wars against Burma, and considerable dynastic turbulence' (1996: 69, 70, see also Chatthip Nartsupha 1986 and Kitahara 1996). This is echoed by Elson in his survey of the peasantry in Southeast Asia in which he describes the region's rural population at the turn of the nineteenth century as, 'in an important sense, still a world to themselves' (1997: 33).[3]

It is important to note that in writing of the 'village', authors are often making reference to several things, although this is not always altogether clear. First, and most obviously, to a physical entity – a cluster or grouping of houses.

---

[2] Scott goes on to state that nowhere is this ideological need to define the past more pressing and urgent than among the poor who have 'collectively created a *remembered village* and a *remembered economy* that serve as an effective ideological backdrop against which to deplore the present' (1985: 178, emphases in original).
[3] Breman, in his challenge to the construction of the traditional Asian village, describes the received wisdom as follows: 'Briefly, the idea that set the fashion . . . postulates the existence in the past of a more or less isolated and self-sufficient local system, which was also autonomous from the administrative and political point of view: the timeless Asian village . . .' (1982: 191).

Plate 4   Wat Xieng Thong, Luang Prabang, Lao PDR. This mirror mosaic shows farmers harvesting and threshing rice. Note the link between this and the picture on page 107.

Second, to a set of social norms, structures and processes that underpin the operation of the village as a social and cultural unit. Third, to an administrative unit with certain institutions (including religious ones) and in particular a village head or chief. And finally, to a loose ideological construct, sometimes encapsulated in the term the rural/village 'idyll'.

Characterising the debate over the nature of rural life in Southeast Asia as sharply polarised, and then presenting the two extremes of the debate clearly harbours the risk of neglecting the views of the majority of scholars that lie somewhere between the two positions. Even those individuals most obviously associated with one camp or the other have found their ideas (predictably) simplified and generalised in the interest of argument. Wolf, for example, views the 'closed corporate peasant community' with which his name is so clearly linked neither as a universal category nor as one half of a closed/open dualism, but as a 'type' or conceptual model (1986: 325–6).[4] As such, he accepts that there are many points on the continuum between closed and open. Bearing in mind the dangers of essentialising the argument in this way, such an approach does have the advantage of clearly revealing the boundaries of the controversy over the nature of the rural past.

On one side of the debate over the nature of the rural past are those scholars and activists who believe that in the pre-colonial era rural settlements were subsistence-oriented, self-reliant, community-based and inward-looking. This is most closely associated with the work of Wolf (1967) and Scott (1976). In this explanatory schema, production was based on local requirements, drew on household or community labour resources, relied on local (or indigenous) technologies, and was calculated with the interests of the 'village community' to the fore. Commercial transactions and interactions were few, if any, and the state only marginally impinged on life and livelihood.[5] They were 'closed corporate peasant communities', often abbreviated to CCPCs. In the opposite camp are those scholars who maintain that commercial interactions were well developed before the modern period, that the influence of the state reached well into the village, and that villages and villagers were outward-looking and dependent on a range of extra-village interactions. These critics sometimes even avoid the use of the word 'village' at all, because of all its attendant associations (in the Western mind) of community, preferring instead to use the less loaded term 'rural settlement' (see Rigg 1994, Hoadley and Gunnarsson 1996). A related area of debate concerns the so-called 'invention' of the village.

It has long been assumed that people in the region have always organised themselves into villages, as if this was the natural and primordial way of doing things.[6] However, scholarship since the 1980s, building on work undertaken in

---

[4] He does admit, though, in a reflective essay written a quarter of a century later 'that the historical perspective embodied in those papers [of 1955 and 1957] now seems overly schematic and not a little naïve' (1986: 326).

[5] Chatthip Nartsupha's (1996) view of the village in Thailand is a good example of such a view.

[6] This critique of village-ism is mirrored in Brow's (1999) discussion of the Ceylonese (Sri Lankan) village and the central role played by the writings of Ananda Coomaraswamy. It became widely accepted that, historically, the Sinhala were a nation of (self-sufficient) villagers peacefully coexisting with one another. This could be traced back over two millennia and was only broken by the advent of British rule in Ceylon

South Asia, has questioned this assumption, and done so in two key ways. To begin with, it has been argued that officials and scholars promoted a European view and interpretation of the operation and structure of the countryside. Thus Kemp (1991: 325) postulates that the village community in Southeast Asia was not a historical construction but a 'theoretical invention arising out of early colonial writings on India and their incorporation into both Marxist and non-Marxist social theory' (1991: 325). In this way the *idea* of the village, and more particularly the village community, was invented to satisfy European mental conceptions. The second way in which the village was invented, is more tangible: it was conjured into existence to satisfy the administrative needs of the colonial authorities and, in Siam/Thailand, the government there. In this manner, the Malaysian *kampong* (Shamsul 1989, Hart 1986), the Indonesian *desa* (Breman 1982, Vickers 1989, Hoadley 1996) and the Thai *muu baan* (Kemp 1988, 1989 and 1991) were imposed on a pre-existing settlement geography, sometimes in a quite arbitrary fashion. The village, as constructed by the central authorities, was then harnessed to extract wealth from the population and the land.

The moral economy in Southeast Asia

James Scott's book *The moral economy of the peasant* (1976) represents the clearest statement of the distinctive quality of peasant society and economy in Southeast Asia.[7] He opens his book by quoting Tawney's vivid likening of China's rural masses to 'a man standing permanently up to his neck in water, so that even a ripple is sufficient to drown him' (1976: 1). Drawing on Chayanov's work, Scott notes that the particular character of the economic life of the peasant is derived from the fact that the peasant family is both a unit of consumption and a unit of production (see also Wong 1987 on this). This results in a 'safety-first' principle where risk avoidance/aversion guides decision-making. Rational economics and issues of profitability take a back seat in this interpretation. But even following such an approach to production there are years when even the most risk-averse peasant faces the spectre of harvest failure and starvation, such are the hazards of peasant life. In such cases, peasants can call on community networks of support and reciprocity – Scott's 'moral economy' – to see them through to the next harvest. The terms that, over time, have come to be associated with such a moral economy provide a summary catalogue exemplifying this perspective on rural life: community, egalitarian, corporate, communal, safe, closed, non-market, mutual assistance, reciprocity, safety-first and sharing.

---

and its attendant commercialism. Villages, in this vision of the past, were 'isolated, self-contained and self-supporting' settlements in which agriculture was 'largely a collective enterprise' and the community self-governing (Brow 1999: 73–4). Furthermore, reciprocity and community-orientation guided village relations. While Coomaraswamy was writing over 90 years ago and may appear, today, 'a rather ineffective and even eccentric figure' (p. 83) his vision of the past can be seen reflected in the ideologies of some village-based development efforts in the country.

[7] Although the book draws on the examples of Lower Burma and Vietnam for much of its supporting material it has come to be viewed as a generic statement on the character of rural society.

This picture of traditional peasant life in Southeast Asia has been contested, most notably by Popkin in his book *The rational peasant* (1979) in which he argues for a political economy approach to the interpretation of peasant choice (see also, Rigg 1994). The emphasis in Popkin's explanatory schema is on the individual rather than the community, and peasants are represented as commercially aware and not averse to embracing money-making opportunities and the pursuit of profit. In a sense this division between the moral economists and political economists is a false one, in at least two senses: first, because the positions of the two groups have never been quite as essentialist as sometimes portrayed; and second, because 'peasants, like capitalists, evidently take risks on some occasions and avoid them on others' (Alexander and Alexander 1982: 599).

Nonetheless, the moral economy perspective is an enduring one and the image that it provokes has become emblematic of the essential character of rural life and society. Few would dispute that rural economies have changed, in some instances beyond all recognition, and yet there is a pervasive sense that beneath the commercial veneer, if only it could be broached, lies the risk-avoiding, community-oriented and moral peasant of old. The power of this image was reflected in the King of Thailand's response to the economic crisis that hit his kingdom, and much of the rest of the region, in 1997 when he called for a return to self-sufficiency and self-reliance (see Box 3.1). But often the view from the countryside was not that the crisis represented an opportunity to rethink development and, perhaps, to rediscover those values lost in the maelstrom of modernisation, but rather that it represented an unwelcome hiatus in the search for modernity. Thus Khun Bunjan, a community leader from Khon Kaen in Thailand's poor Northeastern region, complained in the midst of the economic downturn that 'even our access to schools and health is beginning to disappear', continuing that 'We fear for our children's future' (World Bank 1998: 2). While some scholars, such as Walden Bello (1998), have used the crisis to highlight the inequality, dependency and vulnerability that is, they suggest, part and parcel of global integration, others continue to maintain that Asia's love affair with the market brought enormous benefits to the great majority of its population. Bailey, for example, calls for a higher level of market engagement in the wake of the crisis and flatly rejects the prescriptions of the 'economic troglodytes of [the] Left who seek vindication in people's misery' (2000: 29).

The past as a mirror of the present

Perhaps the most important way in which the past and the present have been conceptualised is as mirror images of one another. Vandergeest has warned against 'theories based on contrasts between the past and the present, in which the past is constructed as that which the present is not' (1991: 423). This can be applied, particularly, to the contrast that is commonly drawn between an egalitarian past and a highly unequal present. Furthermore, it has been argued

Box 3.1   Seeing Thailand through the crisis

Whether or not we are a tiger is unimportant. The important thing is for the economy to be able to support our people. . . . We need to go back so that we can go forward (The King of Thailand, quoted in TDN 1997: 12).

In December 1997, on the occasion of his seventieth birthday and some six months into Thailand's economic crisis, King Bhumibol Adulyadej articulated his New Theory of Agriculture to see the country through its deepest recession since the War (*Bangkok Post* 1998). This was based on the twin notions of economic self-sufficiency (*sethakit porpiang*) and community economy (*sethakit chumchon*) (Chusak Wittayapak 1999: 9).

Politicians fell over themselves to lend their support to the King's message. The Ministry of the Interior said that it would change its rural development strategies to fall into line with the King's speech. It was also said that the eighth National Economic and Social Development Plan would be amended to reflect the King's vision, and academics and NGO workers also quickly lent their support (see Kunsiri Olarikkachat and Ampa Santimatanedol 1997, Falkus and Hewison 1999). Problems in rural areas, and in Thailand more generally, were implicitly linked to the thirst for affluence and a market-orientated economy that 'made the rural sector weak because its resources have been exploited to promote growth in the trade and export sectors' (Kunsiri Olarikkachat 1998). Dr Prawase Wasi, one of Thailand's most respected academics, called for a 'return to modesty and simplicity'. 'This is not the economy that presses for money and that destroys everything including the culture' he said, adding: 'A self-sufficient economy is a moral economy' (quoted in Kunsiri Olarikkachat 1998, see also Chusak Wittayapak 1999: 9). Images of an abundant past were set in opposition to a meagre present and used to justify and support a raft of alternative visions.

The Thai media went out to discover examples of farmers and fisherfolk who reflected the King's message, or who had been wrenched from their traditional livelihoods by modernisation. Tawee Tongthep, a 52 year-old former fisherman, was quoted as saying how before the construction of the Pak Moon Dam in the Northeastern region 'one fishing rod used to be able to feed the whole family'. 'Before we had no bills', he explained, 'now they keep marching in: water, electricity, children's tuition, medical expenses . . .' (quoted in Vasana Chinvarakorn 1998).

Academics have described this recent attempt to challenge the nature of Thailand's development as a 'localism discourse' (see Hewison 1999: 10). While the King's vision has been presented as new in some quarters it can be directly linked to the country's rather more venerable Community Development Perspective and from here back in time to some of the tenets of Buddhism. As Hewison writes: '"Modern agriculture" is identified as having destroyed the assumed abundance of the past' (1999: 12). But it is not just a case of commercialism undermining agriculture; modernisation has also eroded, in the parlance of this localism discourse, the generosity, mutual assistance and morality that characterised traditional rural society. A rich seam of populist rhetoric about the past infuses this vision of a more desirable future and, as Hewison argues, it provides no politically or economically sound alternative to the existing strategy of global integration (1999: 22).

Plate 5 Hanks of black (*nasi hitam*) and white rice waiting to be stored in a rice barn in the highland Toraja area of South Sulawesi, Indonesia.

that this profound shift from equality to inequality has been driven by the operation of the market and the intrusion of the state into village life and affairs.

The literature on the nature of pre-colonial village life in Southeast Asia is richest on Java. Clifford Geertz in his seminal book *Agricultural involution: the processes of ecological change in Indonesia* (1963) set out the case for pre-colonial uniformity. However since then, detailed historical work by Boomgaard (1989), Breman (1982), Knight (1982), Hart (1986), Alexander and Alexander (1982), Hüsken (1981), Svensson (1991), White (1983) and Carey (1986) have shown, in their different ways, that villages in Java were complex and stratified, where the interests of villagers (the 'community') did not coincide and, indeed, were sometimes sharply at odds. Hart (1986: 20), Carey (1986: 86) and Boomgaard (1989: 67) all note a significant divide between large landowning classes and landless peasants.[8] Furthermore, this divide existed well before the full-scale colonial exploitation of Java. Indeed, the historical evidence shows that peasants on Java were producing excess rice for sale well before the first millennium AD (Boomgaard 1991a: 16).

Two important strands in this reassessment of the nature of rural life – in Java and more widely in the region – concern the operation of the market and the state. As noted above, scholars had hitherto tended to imagine that the pre-colonial village was inward-looking and largely self-reliant, isolated from the market and the state – an autonomous Little Republic. While there can be little doubt that the degree of market and state integration is far greater today than ever before, historians have nonetheless been surprised at the level of interaction even in the pre-colonial period. Anthony Reid, in two important books (1988 and 1993), offers a sustained argument that during the period 1450–1680, before the colonial era, Southeast Asia enjoyed an 'age of commerce'. During this 230-year period Southeast Asia became a region of intense commercial activity and state centralisation where 'large proportions of the population became dependent on international trade for their livelihood, their clothing and everyday necessities, and even their food' (1993: 327). This age of commerce ended in the seventeenth century as Asia was eclipsed by the emergent European powers, but the region did not atomise and commercial activity was not snuffed out.

Bowie (1992) and Koizumi (1992) in their work on Thailand and Elson (1986) and White (1991) on Java describe villages that are differentiated, dynamic, and engaged in market exchange. In Northeastern Thailand, a region remote from the centre, the payment of the *suai* head tax in cash rather than kind had become the norm by the middle of the nineteenth century (Koizumi 1992). Gold, silver and baser coinage were in circulation in rural Java by the ninth century (Boomgaard 1991b: 294) and the island was highly monetised by the early nineteenth century, as attested by the prevalence of cash wage labouring

---

[8] 'Within rural society, the pattern of access to land was highly differentiated . . . and a class of dependent landless households existed long before the full-scale colonial exploitation of Java' (Hart 1986: 20). '. . . It is clear that the structure of Javanese peasant society, at least until the first decades of the nineteenth century, gave important advantages to the group of "landowning" peasants who could prosper independently by drawing on the labour services of a resident work-force of dependents and landless peasants' (Carey 1986: 86).

(White 1991). In the Batak area of inland Sumatra coinage, though scarce, was widespread before the imposition of Dutch authority, and was used for bride-wealth and for land purchases (Sherman 1990: 46). And in Northern Thailand the market in textiles was sufficiently well developed by the nineteenth century for Bowie to conclude that 'this examination . . . reveals a society with a complex division of labor, serious class stratification, dire poverty, a wide-ranging trade network, and an unanticipated dynamism' (1992: 819).

As with market exchange, so with state intervention: there was more of it than was hitherto imagined. Vandergeest (1991: 440) remarks on the relations of exploitation that characterised nineteenth-century Thailand; Elson (1986: 65–6) writes of the degree to which rural communities in early nineteenth-century Java were far from being isolated and independent; and Terweil (1989: 251) notes, with surprise, the 'efficacy of the 19th century government apparatus' in Thailand where even 'living at some distance from Bangkok did not place a person beyond the reach of government officials'. Messages from officials in Bangkok 'would directly affect the lives of farmers they would never personally see'. There are periods in Java's history (the twelfth and fourteenth centuries, for example) when the state's centralising tendencies and taxation were well developed (Boomgaard 1991b: 293–4).

There would seem to be mounting historical evidence to suggest three things about pre-modern rural Southeast Asia. Firstly, it appears that the degree of market integration and exchange, and social and economic differentiation, was somewhat greater than hitherto assumed. The image of the region as 'immovably static', one in which 'capitalism did not exist and which resisted change until the 19th century' is, as Day has maintained, a fiction (1986: 1). Drawing on epigraphic evidence from early Java, Christie writes that the 'Javanese state, its constituent communities, and their relationships changed constantly' (1994: 40). Secondly, it seems that there was considerable spatial and temporal variation in the degree of integration and differentiation.[9] Some rural populations were less engaged in market exchange and inhabited marginal spaces comparatively remote from the centre. Political and economic turmoil also occasionally promoted greater self-reliance, encouraging villages with considerable commercial activity to retreat into subsistence. So the argument that pre-modern Southeast Asia was more commercialised, more differentiated, and more vital than imagined is a generalisation which often falls foul of the minutiae of change – or the 'eddies and currents of specific events', as Reid puts it (1993: 327). And thirdly, evidence suggests that the contrast highlighted above is falsely drawn at least to the degree that the divisions are not exclusive. Community and commerce are not irreconcilable; production for subsistence and production for market can occur side by side; and village autonomy and state intrusion are not necessarily antithetical. Indeed, some of the key elements of communal rural life in Java were colonial creations. This applies, for example, to the *sawah desa* or communal lands (Boomgaard 1991b: 293).

---

[9] See, for example, Rambo's (1977) discussion of the 'closed' and 'open' villages of Northern and Southern Vietnam.

The state in the village

It has already been noted that the role of the state in villagers' lives, even before the height of the colonial period in Southeast Asia, was probably more intensive and more profound than hitherto believed. But, like many such issues, there is an important question of degree to be considered. It seems reasonable to write that as the colonial powers tightened their hold on the region so villagers were further drawn into the embrace of the state (and the market). This reached a higher plane still after independence, and in the subsequent development decades. We might consider the modern state, at the turn of the twenty-first century, to be (in that unfortunate but evocative modern phrase) an 'in your face' state.

Hirsch's (1989, 1990a, 1990b, 1992) work on Thailand represents one of the most insightful series of investigations into the evolving links between rural people, the state and the market in Southeast Asia. In Hirsch's view (1989b: 36), there has been a parallel, but in his view contradictory, process of state infiltration into the village (characterised as domination) and villager integration into the state (which is more akin to the extension of civil society). But it is important to see the state allied to capital in this process of infiltration and domination. State and capital have operated in tandem and, often, in collaboration to create new and more productive spaces (social and spatial) for the extraction of surplus from the countryside.

Hirsch uses the case study of Ban Mai in Thailand's western frontier province of Uthai Thani to make his case. Prior to the late 1970s, Ban Mai corresponded, apparently, with the image of the closed village community described earlier in this chapter. Government officials were virtually non-existent, there was no formal village head, no school, the police were rarely seen and, overall, contact with the state was 'minimal' (1989b: 39–40 and see Hirsch 1990a: 71–4, 79–80). Beginning in the late 1970s the state rural development machine began to impose itself on Ban Mai in the shape of new institutions and administrative structures, and new development opportunities from institutional credit through to various community development projects. While village-level institutions like the Tambon [sub-district] Council and Village Committee are evidence of village-level democracy, they are also the means by which the state can infiltrate and influence village affairs. And, of course, along with the state came the market. Furthermore, influential villagers were able to dominate and control village-level institutions and thereby gain more effective access to the largesse of the state. This, in turn, provided advantageous access to the money-making opportunities offered by the interests of capital. As Hirsch explains of Ban Mai, 'clearly this is a village that has been articulated rapidly by state and capital after an earlier period of isolation and autarchy' (1989b: 42).

Hirsch provides a powerful image of a past in which the traditional village leader, the *khon to*, worked in the interests of the community as a whole, and a present where the new official leader, the *phuu yai baan*, became the 'mouthpiece of the state in the village' (1989b: 53, see also Hirsch 1990a: 203–7) and wealthy to boot through illegal timber and land sales (the two being closely linked). Critical to Hirsch's analysis is a fundamental shift from the village as largely independent of the state to a situation today where the village has 'ceased to exist independently of the state, either in a geographical or an institutional sense' (1989b: 54). The village has become (a part of) the state. Without the state, there is no village.

It is possible to challenge, in its generalities if not its specifics, Hirsch's image of capitalist relations of production infiltrating 'isolated subsistence communities' (1989b: 37). Vandergeest, for example, in contesting the 'pervasive assumption . . . that development, markets and state-building are undermining autonomous communities' decries the tendency to construct dualisms that fail critically to challenge historical assumptions. Indeed, Vandergeest goes so far as to suggest that villagers in Thailand are 'as close [today] as they have ever been to an autonomous, community-based, form of development' (1991: 434). While the tenor of this chapter has been to support Vandergeest's position, the picture that Hirsch portrays of an intensification of the articulation of villagers, state and capital is hard to resist.

A similar story emerges in work on Indonesia, and especially on Java, where commentators seem agreed that under the New Order of former President Suharto (1967–1998) the state strongly and effectively penetrated the countryside (see Pincus 1996, Hill 1996, Fox 1991 and 1993). A significant element of this was the improvement of roads which brought the state, arm in arm with the market, to the village (see the discussion of this in Chapter 4). But also critical, as in Thailand, were various other elements of the modernisation project ranging from the New Order's massive rice intensification programme through to its highly effective family planning programme and the state's role in controlling village political life.[10]

Work in Malaysia tends to recount a similar intensification of state–village and market–village relations, although the effects are usually not interpreted as being so pernicious as in the cases of Thailand and Indonesia. Scott (1985), for example, in Sedaka describes in great detail the various agencies of the state that impact on the inhabitants' lives from the Farmers' Association through to UMNO, the ruling Malay political party. De Koninck too, in his longitudinal study of the villages of Matang Pinang and Paya Keladi, also in the Muda area of Kedah state, notes the degree to which the state has intervened in people's lives. This extends from the provision of social and physical infrastructure, through to the state's role as a major employer of local people and its involvement in almost every stage of the rice production cycle. The state subsidises and distributes inputs, supports the rice price and acts, simultaneously, as middleman, benefactor and entrepreneur (De Koninck 1992: 149–50). De Koninck sums up the state's role in people's lives and livelihoods in Kedah by writing of its 'omnipresence in the countryside' and 'its close monitoring of just about every aspect of peasant agriculture' (1992: 152).[11]

---

[10] See Cederroth's (1995) account of this in Bantur, a village near Malang in East Java.

[11] Both Scott and De Koninck conducted their fieldwork in villages included within the area of the Muda irrigation scheme. It is worth noting that this scheme represents Malaysia's most ambitious and intensive agricultural development effort to date (see Jirström 1996). A clear and close role for the state is therefore not altogether surprising.

Plate 6 Rice terraces, Central Java. It is images like this, of an apparently unchanging bucolic landscape that lend credence to the popular notion that rural areas of Southeast Asia are relatively unchanged and unchanging.

The history of modernity: peasant to post-peasant

Strongly associated with the image of the village and rural areas outlined above is the notion that modernity 'arrived' or colonised rural Southeast Asia around about 1945. Escobar, notably, has equated the arrival of development (or the culture of modernity) in the Third World with the arrival of the Allies in Europe during D-Day. Not only are the dates the same, but so too is the image of invasion by an overwhelming force against a resistant, but ultimately powerless, opposition. Escobar writes of how scholars and politicians began to fulminate about the hunger, disease, squalor and misery which was regarded as endemic in the Third World, and writes of the 'War Against Poverty' which the post-war era offered up as the new battle facing the Free World (1995: 21).

But this image, tempting and enticing though it may be, flies in the face of historical research. To begin with, there is the implicit assumption that before 1945 (or 1870 or 1900 or the seventeenth century – views vary on this point) there was no modernity. Not only is the implied dualistic modern/traditional dichotomy unsustainable but there is also ample evidence from across Southeast Asia that 'the desire to embrace the new was not new' (Vickers 1996: 9). Bali, for example, was not a closed culture before modernity in the shape of the Dutch brushed its shores. Trade, commerce and other forms of social and economic interaction were common and entrenched (Vickers 1996). The terms 'modern' and 'traditional', and their application to change in the region are deeply problematic. This links back to the image of Southeast Asia as unchanging and ahistorical before the colonial powers and colonial minds imposed a history on the region. In this view of things, the modern is clearly situated in 'Europe' (Day 1986: 11). Day argues that historians have been at pains to ensure that the 'modernity of the modern and the traditionalism of tradition' are maintained and he suggests that one way to push back the history of modernity in an island like Java is to 'argue for a fully contextualised definition of what the modern is in a Javanese sense . . .' (1986: 26). Like Vickers (1997: 177–81, 283), he finds Anderson's view that colonialism formed a critical break between the traditional and the modern problematic in both a historical and a categorical sense. Discontinuities are rarely as sharp as portrayed, and there is something more than just the modern and the traditional (see Cowan and Shenton 1996).

Elson, in his ambitious book tracing the history of the Southeast Asian peasantry from 1800 to the 1990s, defines his subjects as 'rural groupings of people whose primary orientation, both economic and social, was towards a broad participation in simple, relatively unspecialised, household-based, subsistence agricultural production . . .' (1997: xix). But at the same time as he offers a loose definition of the peasant, Elson agrees with Shanin (1990) in emphasising that the idea of the peasantry is a process. The peasant of one era, it seems, is not the peasant of another. Peasants, therefore, are contextually defined, both historically and socially (Shanin 1990: 51, Shanin 1987).

For Elson, the great age of the Southeast Asian peasantry spanned the years from the middle of the nineteenth century through to the early decades of the twentieth century when 'peasant labour and production sustained, as never before, the economies of the region' (1997: xxi). Producing vast quantities of rice, rubber, coffee, tobacco, pepper, coffee and sugar, the peasants of Southeast Asia found themselves at the core of the colonial economic project in the region. With a strengthening and more effective colonial state apparatus, the Southeast Asian peasant was drawn, however reluctantly, into the embrace of the global economy.

To use the term 'peasant' to describe rural dwellers in Southeast Asia today would be odd. As the later chapters of this book will demonstrate, the modern peasant is not a peasant at all unless, that is, the word is redefined to fit with the reality. He or she may have a home in the countryside, may continue to grow rice, and may inhabit a village and a landscape which maintain a visual link with the past. But events have consigned the peasantry, as a social formation, to history. The peasant, to use Baker's phrase, has become a 'post-peasant' (1999: 4–5).

Chapter 4

Imagining the future: aspirations, mobility and modernity

'Staying in the village has no future, nothing.' (An Akha woman who had left her village in the hills for the city of Chiang Mai, quoted in Toyota 1999: 273).

Introduction

In addressing the twin issues of aspirations and mobility, this chapter is concerned with two of the critical factors that propel change in rural communities – and entice, encourage and permit people to become modern. The utility of goods is culturally determined and as aspirations escalate so consumption patterns change. The determination of needs is embedded in cultural practice and social norms. Adam Smith expressed this with his usual clarity when he wrote over two centuries ago that 'necessities' are 'not only the commodities which are indispensably necessary for the support of life, but whatever the custom of the country renders it indecent for creditable people . . . to be without' (Smith 1776). Consumption patterns are, in turn, closely related to issues of communication – spatial and social. People and communities communicate consumption preferences, establishing by doing so accepted patterns of behaviour. At the same time, communication in the form of mobility allows the market and state to intrude into rural areas, and rural people and their products to infiltrate into the wider arena.

Taken together, these forces help to explain why rural society is becoming more commodified and rural livelihoods increasingly integrated into wider systems of exchange. Interpenetration – of ideas, desires, goods, people, information and money – has served to challenge the enduring belief that there is a distinctive rural character that can be offered up as somehow different. In some instances traditional cultural norms and new consumption preferences exist side by side. In other cases there is an emergent tension between the modern and the traditional, while at other times the traditional may be snuffed out by an overpowering modernity (see the section opposite on consumer cultures).

Table 4.1   Measuring necessity in Vietnam

Item or service	Number of respondents who consider this item a necessity	% of respondents
Doctor visiting a house when sick	418	99.5
All children studying to level two	418	99.5
One sao (500m²) of land per person	418	99.5
Buffalo or cow	415	98.8
Three meals a day	414	98.6
Thick blanket	413	98.3
Wooden rice chest	412	98.1
Concrete rice drying yard	412	98.1
Well with well head	411	97.9
Bicycle	410	97.6
Electric light	408	97.1
Pesticide pump	399	95.0
Livestock vaccination	391	93.1
New set of clothes each year	388	92.4
Stone-built toilet	386	91.9
Electric fan	350	83.3
Meat once a week	336	80.0
Access to VBA loans	325	77.4
Stone-built house	322	76.7
Bathroom	312	74.3
Table and chairs made of good wood	188	44.8
Watch	175	41.7
Radio	135	32.1
TV	88	21.0
Two compartment wooden wardrobe	78	18.6
Motorbike	32	7.6

Source: Davies and Smith 1998: 5

Tracking necessity and the rise of consumer cultures

One perspective on the dynamism that is embedded in notions and conceptions of necessity is provided by ActionAid's study conducted towards the end of 1997 among 420 households in Can Loc, a district in Ha Tinh province in north Vietnam (Davies and Smith 1998). The study attempted to gauge whether respondents regarded a list of items as necessities. The results show that 99.5 per cent considered education of children up to level two as a necessity, 97 per cent considered electric light as such, 83.3 per cent an electric fan, 41.7 per cent a watch, and 21 per cent a television (Table 4.1). In one regard the study is reassuring in confirming a 'common sense' view of necessity. Those services, amenities and products at the top of the table are not luxury consumer goods but items linked to issues of health, education, nutrition, sanitation and factors of production such as land and livestock. But in another regard it may be considered surprising that in this poor area of a poor province in a poor country over one-fifth of respondents thought a television a 'necessity', one-third a radio, and over 40 per cent a watch.

Evidence from other, wealthier areas of Southeast Asia clearly indicates that as Vietnam modernises and its population becomes wealthier so we can expect that the perceptions of necessity will shift. In Thailand, for example, studies reveal that in most rural areas TVs are now regarded as necessities (Baker 1999: 4) while in some places motorbikes and pick-up trucks are viewed as such (Singhanetra-Renard 1999: 80).[1] Preston's (1998) longitudinal study of two villages in the upland Cordillera of northern Luzon in the Philippines likewise highlights the sometimes bewildering pace of change – and the clashing visual images that can result. Bent old women in home-woven skirts alongside men wearing T-shirts and g-strings and women in modern designer clothes listening to international rock music; modern-style homes of brick and galvanised iron alongside traditional granaries (Preston 1998: 380–1).

There are people, communities and places in Southeast Asia which do not accord with this image. But they are surprisingly few.[2] Consumerism has arrived! For Hefner (1990), working in the Tengger highlands of East Java, there are 'few developments . . . more striking, or more symptomatic of the changing spirit of the age, than recent shifts in conspicuous consumption' (Hefner 1990: 182). Education, the experience of cities and urban life, and the consumerist messages conveyed by television, radio and newspapers and magazines, all contribute, brick by brick, to an emerging culture that can be best described as modernist and consumerist. Funahashi writes of the 'stylish and coveted lifestyles portrayed on television' that entice young people in Yasothon province in Thailand's poor Northeast away to the city (1996: 108). The pressure to be modern is hard to resist. To rural producers 'the images of urban wealth and commodified progress that pervade Thai popular culture pose models of consumption and social status that, however difficult to achieve, are impossible to ignore' (Mills 1997: 42). The Thai rock band, Carabao, in their song *Made in Thailand*, written during the early years of the kingdom's economic boom when it was the fastest-growing economy in the world, sang of the consumer pressures on the young: 'Afraid to lose face, worried their taste isn't modern enough. . . . Attach a brand label . . . It'll sell like hot cakes, fetch a good price. . . . We can tell ourselves it's foreign made' (Pasuk Phongpaichit and Baker 1998: 155).

Particularly striking is the speed with which a luxury can be transformed into a necessity. In Klong Ban Pho in Central Thailand's Chachoengsao province, the pick-up truck made the transition from 'status symbol' to 'necessity'

[1] Rural TV ownership increased from 30 per cent to 70 per cent over the course of the 1980s and by the mid-1990s exceeded 90 per cent. Motorcycle ownership saw a similar rate of increase. In 1992 around one half of rural families owned a motorcycle; the figure for 1996 was 71 per cent (Baker 1999: 4).

[2] Occasionally individuals buck the trend. A 24-year-old male Chiang Mai-based Akha informant explained to Mika Toyota: 'I worked hard in order to buy my clothes, a cassette tape recorder, bike, television and refrigerator, etc. Then, I realised that the desire for consumption does not stop. If I bought a CD player, I would want CDs and then my desire would be for a camera, a car, computer, etc. Because money would never be enough, no matter how much you earn, I am tired of it' (1999: 311).

in the space of just five years (Chantana Banpasirichote 1993: 58). Indeed, Klong Ban Pho's recent history offers a telling vignette of the changes in consumption patterns that are sweeping through the region. A surfaced road was built to Klong Ban Pho in 1975. It was not until 1987 that electricity became available to villagers. At around the same time people began to engage in factory work and diversified into new farm activities, most notably intensive shrimp farming. By the early 1990s factory work had become the preferred livelihood option, especially among the young, and shrimp farming was in decline. It was also at this time that consumerism made its mark. The motorbike and pick-up truck had become key markers of success, it was common social practice to visit the newly built shopping centres in the provincial capital of Chaochoengsao some 17 kilometres away, and modern fashions dominated the village. From being a traditional, subsistence-oriented, agricultural community without surfaced road or electricity in 1975, Klong Ban Pho had been transformed in some 15 years into a village of enthusiastic consumers with considerable disposable income and a wish to avoid agriculture.

The tensions at the interface of modernity and tradition

Unsurprisingly, there is a tension between advancing modernities and resistant traditions. This is reflected most clearly in the position of young women. Work from the Philippines (Tacoli 1996, Kelly 1999b: 72–3), Malaysia (Hart 1992), Thailand (Mills 1997, Bencha Yoddumnern-Attig *et al.* 1992, Jones and Tieng Pardthaisong 1999: 46) and Indonesia (Wolf 1990 and 1992) illustrates that daughters are under considerable moral obligation to support their families. This is often articulated in terms of having to 'repay' their parents for bringing them up (Chant 1998: 12) and is reflected in the title of Tacoli's (1996) paper, migrating 'for the sake of the family'. The familial obligations of sons, it seems, are considerably less.[3] Hart (1992) notes that young men in her study area in the Muda region of Peninsular Malaysia sometimes made a financial contribution to their upkeep but that they were under no compulsion to do so. Neither were they required to contribute labour to the household. One mother in Sungai Gajah, when asked what her son did angrily replied *Dia kerja makan* – 'He works at eating' (Hart 1992: 813).

Mills (1997), in her study from Thailand, writes of the conflicts between 'modern women' and 'dutiful daughters'. Increasingly, young women are being drawn into modern employment partly due to a need to generate income but also due to a desire to be modern or *than samai*. Modernity, in these terms, is multifaceted. It means, for example: incorporation into the modern economy; engagement in a 'clean', modern occupation; living in the city; wearing modern clothes; having white skin; financial autonomy; and comparative freedom of action (see also Arghiros 1998 on the attractions of a factory life). Kelly explains the consumption decisions of young people in Cavite, the Philippines as

---

[3] Although Kelly's survey of 146 high school leavers in Cavite south of Manila reveals that of those expressing an interest in working abroad 38 per cent of males as against 34 per cent of females explained this in terms of helping parents and siblings (1999b: 73).

reflecting 'the adoption of a locally hybridised "global" culture'. He continues: 'This is inscribed in the dress and appearance of young people – earrings, clothing styles, tattooing, and the use of skin-lightening soaps to achieve a paler, more "western" appearance' (Kelly 1999b: 67).

However, engagement in modern, non-farm employment also permits a woman to be a dutiful daughter by supporting her family through regular remittances and the giving of gifts. Indeed, engagement in the modern economy often requires young women to behave in this way (as noted above, there is a moral imperative at work). This, in turn, creates a tension between the desire to be modern and the necessity to be dutiful. Commodity consumption under-scores and motivates the movement of young women into urban-based factory work, while powerful cultural discourses mould an image of the dutiful daughter that ensures a significant proportion of income earned is diverted back to the family.

In some instances – and the evidence comes particularly from Thailand – this sense of duty and obligation supports and locally justifies work in the sex industry. Jones and Tieng Pardthaisong (1999) for example write of the case of three women who worked for two years as commercial sex workers in Japan. In each case they regarded their experience abroad as 'successful' both in material terms (the income they earned allowed two of them to build new houses and all three to live comfortably) and because it permitted them to meet their familial obligations. One woman explained:

> I am very proud to be able to make my parents happy . . . In the past, the neck-lace I wore was very cheap. I used to look after buffaloes, and people looked down on me. But if I have money, everybody will respect me . . . Now I do rice farming – just for fun, but also I am proud of it (Jones and Tieng Pardthaisong 1999: 46).[4]

This neatly encapsulates the twin pressures that arise from a daughter's obligation to her parents and the association of wealth with respect. Remitting money, bringing presents, perhaps building a house for her parents, demonstrates 'a daughter's virtue, her willingness to sacrifice herself, her gratitude to her parents and, more importantly, her success' (Sanitsuda Ekachai 1990: 171). That the ethos of modernity, coupled with a sense of duty, can justify – indeed promote – even work in the commercial sex industry is particularly striking but, for Thailand at least, it is not unusual (see Sanitsuda Ekachai 1990: 169–70, Pasuk Phongpaichit 1984, and Rigg 1997: 131–4).

A similar theme of tension between duty and modernity is evident in Elmhirst's work in the Indonesian province of Lampung in south Sumatra (1998a and 1998b) and Tacoli's work on Philippine international migrants in

---

[4] Sanitsuda Ekachai, writing about a village in Phayao province in Northern Thailand, makes the same point: 'Why do they [the villagers] let their girls, as young as 13, be sexually violated at the whim of strangers? Is it hunger? Poverty? Or is it greed? All such questions have to be swallowed when coming face-to-face with Pon Chaitep, a peasant who let his daughter go to work as a prostitute in Sungai Kolok, a town on the Malaysian border, in exchange for 15,000 baht [about US$600] . . . "I didn't sell my daughter," mumbles the father of eight apologetically. "She saw me suffer. She saw the family suffer. And she wanted to help"' (1990: 169–70).

Rome. Tacoli notes that female migrants remit considerably larger sums of money to their families than male migrants, and on a more regular basis. Eldest daughters, particularly, are under a considerable familial obligation to provide financial support for their families and particularly to support the education of their younger siblings. This demand makes it difficult, if not impossible, for daughters to fulfil their filial duty while also meeting their personal savings needs (1996: 21–22).[5]

Consuming the rural: the colonisation of the countryside

Research on rural change in the developed world has emphasised the degree to which rural areas are being infiltrated by new rural classes (see Chapter 2). These classes are rarely engaged in agriculture and other traditional rural occupations. Indeed, they are detached from such pursuits, both experientially and mentally. Instead, their presence in the countryside is propelled by considerations which have more to do with consumption than with production. The countryside is exploited as an area with certain desirable characteristics embracing, for example, environmental quality and leisure and recreational opportunities. The search for the elusive rural idyll is highlighted in the developed world literature as important in enticing people into the countryside.[6]

In Southeast Asia – and much of the rest of the developing world – work of this type is thinly represented. Analyses of rural change are still dominated by issues of production, rather than questions of consumption. Nonetheless, there is evidence that some areas are making a similar transition from being zones of production, to becoming landscapes of consumption. Mae Sa, in the Chiang Mai Valley of Northern Thailand, is just such an area.

Mae Sa has been studied by Anchalee Singhanetra-Renard over some 22 years. When Singhanetra-Renard (1999) began her work there in the mid-1970s, Mae Sa was an archetypal Northern Thai farming community. The majority of households cultivated wet rice to meet their subsistence needs and supplemented this with vegetable production on sandbars in the Ping River, some upland crop cultivation on higher land, and a modicum of off-farm work. At this point in the social and economic evolution of Mae Sa, the focus of life and livelihood was very much on the village and on agriculture. However, with its location just 13 kilometres north of the city of Chiang Mai – Northern Thailand's unofficial capital – it was inevitable that Mae Sa would be profoundly affected by the growth and expansion of the city. In the late 1970s the first land agents came to the village to purchase the vegetable gardens near the Ping River. But it was not until the mid-1980s that land buying on a 'massive' scale occurred. In 1993 the last village rice field was sold, by which time Mae Sa could no longer be described as an agricultural community, even a partial

---

[5] Tacoli also notes that female migrants in Rome are under considerable pressure from their parents back home not to marry during their time away because this would mean that their primary obligation would shift from their family to their husband. As such, they would no longer be able to act as dutiful daughters, but would be obliged instead to become dutiful wives (1996: 22).

[6] There are also forces which encourage people to leave urban and suburban areas, from pollution to crime.

Plate 7 Vietnam's future – school's out.

one. By 1996, with the exception of a handful of local shopkeepers, everyone living in the village commuted to work and agriculture had been almost entirely displaced from the surrounding land. 'Within a 15-kilometre radius of the village', Singhanetra-Renard writes, 'there are golf courses, reservoirs, and elephant shows; orchid, butterfly and snake farms; restaurants, five-star hotels, karaoke bars, brothels, massage parlours and resorts . . .' (1999: 77).

The politics of modernity

There has been a tendency to view the spread of consumer cultures across Southeast Asia as just another facet of the operation of the market. Yet it is evident that the state has been actively and intimately involved in the process. It is also clear that the bringing of modernity to the rural masses is an intensely political process.

The intertwining of politics with modernity and notions of progress is probably tightest in Indonesia. The Indonesian word for development, *pembangunan*, has powerful connotations of planning and control and emphasises 'the need for guidance by those with power and knowledge, in this case the government officials who elaborated the notion in the first place' (Hobart, 1993: 7 and see Rigg *et al.* 1999). Notions of progress or *kemajuan* also permeate official discourse. During the course of his 32-year presidency (1967–1998), Suharto appropriated both the word and the ideology that underpins it. In time he became popularly known as *Bapak Pembangunan*, or the Father of Development, and ended up occupying a 'central role in the edifice of Indonesian development [thus creating] an explicit link between state legitimacy and modernization' (Rigg *et al.* 1999: 586). This was evident both in the higher reaches of government and also at the local level. Cederroth, in his account of Golkar meetings in a village in East Java in the run-up to the parliamentary elections of 1987 writes that 'a concept to which all the speakers returned again and again was Pembangunan (Development) [which] was said to have taken place in almost all spheres of life under the wise and careful guidance of Suharto . . .' (1995: 34, see also Klopfer 1994).

As well as its associations of state control, the other side of the *pembangunan* coin is modernity. In general terms, it means the essentials of progress or *maju*: electricity, surfaced roads, modern health care, new rice technology, education, and so forth. But it also embodies the culture of consumerism. Alison Murray, in her account of kampung life in Jakarta writes that '[a] vinyl three-piece suite, coffee table and lacquered buffet are considered essential symbols of social standing, and it was considered shameful that one of the RT [neighbourhood] leaders in RW [community] 'B' had no lounge suite and had to entertain visitors at his father-in-law's house' (Murray 1991: 37). Elmhirst notes a similar tendency for young women from Lampung who have worked away from home in the textile factories of Tengarang, outside Jakarta, to spend large sums of money acquiring certain essential consumer goods (Elmhirst 1998b). Importantly, there are strong value judgements attached to the achievement of progress. To be modern is to be advanced; to be *belum maju* ('lacking progress') or *masih di belakang* ('backward') is to be deplored (see Elmhirst 1997). Antlöv takes this further to suggest that *pembangunan* ideology is characterised by the philosophy of 'betting on the strong' (Antlöv 1995). Villagers are encouraged to become modern and have been encouraged into thinking of modernity as necessarily good.

Hettne, in accounting for the failure of alternative visions of development (small-scale, need-oriented, community-based, endogenous, appropriate, environmentally sustainable and benign) to take root in many countries of the South points, in explanation, to the influence of unequal power relations, corruption and the lack of representativeness of governments (1990: 155 and 1995: 161–3). But, while there is a strong sense in which the couplet development/modernity is orchestrated and disseminated by the state, it would be wrong to see development as modernisation being imposed on a dissenting, unwilling or impercipient population. The modernisation ethic has been internalised and is, today, a largely uncontested goal. In Bali, 'the fascination with being modern goes back beyond the origins of the present government [of Suharto], is shared on virtually all levels of society, and is common to both the Left and the Right' (Vickers 1996: 1). This sentiment also rings true for many other areas and countries in the region. Being modern is not questioned, although how to achieve modernity may be.

Though I have argued that all people in Southeast Asia – with few exceptions – wish to be modern, this is not to say that modernity is the imposition/adoption of a homogenous experience which has its source in the West.

Plate 8 Children in school, Northern Thailand. Every one of these children in their final year of primary level schooling expressed a desire to continue with their education. Although all came from rural backgrounds not one wished to be a farmer or work in agriculture.

Modernity is culturally shaped so that when a Thai talks or thinks of *than samay* or an Indonesian, *moderen* (both meaning modernity) they are doing so through a cultural lens. The words do carry a recognisable and substantial element of Western baggage that an American or European would quickly recognise. But they are also different. So too with cultures of consumption. They are familiar and recognisable as embodying elements from a notional global consumption discourse, but they are far from being identical.

Education, education, education

Let me start with a generalisation: in the past children, and especially sons, were expected to work the land; today they are expected to go to school, study and acquire the necessary and appropriate education and skills so that they can abandon the land. It is the duty and responsibility of parents to assist in this quest.

Kelly conducted a survey of 123 school leavers in Tanza's High School in the province of Cavite south of Manila. Not a single respondent expressed an interest in working in farming. Instead 20 per cent were intent on working in the Cavite Export Processing Zone after they had graduated, 10 per cent in Manila, 5 per cent wished to work abroad, while 65 per cent expressed a desire to continue with their studies (1999b: 70). 'For the vast majority, working in a factory or office or going overseas are their preferred options. . . . This applies even to those whose parents are farmers . . .' (1999a: 297). The same survey recorded students' perceptions of farming as an occupation. Over three-quarters expressed their views in negative terms ('hard work', 'boring', 'low income') while the 20 per cent who were more positive emphasised the benefits that working in farming brings to the country and to their parents (but not to themselves) (1999b: 71).

Among the Toba Batak of North Sumatra the importance placed on education can be traced back to the influence of the first missionaries at the end of the nineteenth century (Rodenburg 1997: 143, Bruner 1961: 510–12).[7] Indeed, many of the issues highlighted in this section as being of fairly recent origin have, in some instances, a comparatively long history. Bruner believes that the effect of several decades of German missionary education, part-funded by the Netherland Indies government, had already had a profound effect on Batak aspirations even before the Japanese occupation in 1942. 'Education', he writes, 'was the "golden plough", the means of escaping from the drudgery of work in the rice field and the monotony of village life' (Bruner 1961: 511). In terms of the Southeast Asian picture as a whole, the Batak experience can be counted unusual. But this quote, referring to an era now several decades in the past, highlights three of the key issues underpinning contemporary rural change in Southeast Asia: the emphasis placed on modern education; the desire to escape the drudgery of farming; and the wish to flee from the village.

For most areas and people the emphasis on acquiring an education really began to take root after 1945, intensifying with each decade. Hefner's work on the Tengger Highlands of East Java in Indonesia illustrates the changing perception of the value of education in the years since independence. Until the 1970s, he maintains, parents endeavoured to keep their children at home, in their natal villages. An education beyond the sixth grade required a move to the lowlands (there was no middle school in the area until 1978–79). This was resisted on cultural grounds (it would mean mountain Hindus sending their children to Islam-dominated lowland schools and communities) and for apparently good economic reasons. As villagers explained to Hefner, the '. . . income from a good-sized farm was better than that of a lower-level government official . . .' (1990: 178). But this perspective on education changed fundamentally during the course of the 1970s, for a range of reasons. To begin with, education provision improved dramatically as local schools were built and staffed. Second, the culture of modernity, of which education was a central component, began to affect decision-making. And third, the perceived economic value of an education (and, by extension, non-farm employment) rose while that of agriculture declined. In the 1960s none of the upland elite wished to see their children leave the village to pursue their studies; by the 1980s many did so (Hefner 1990: 179). Toyota (1999) reports a similar chain of events among the hill peoples of Northern Thailand. In the mid-1980s it

---

[7] In 1920 the literacy rate among the Toba Batak was 5.9 per cent; ten years later, in 1930, it had more than doubled to 12.4 per cent (Rodenburg 1997: 144).

Table 4.2 Percentage of males and females aged 15–19 enrolled in school in Indonesia, Philippines and Thailand (1950–90)

| | Indonesia | | Malaysia | | Myanmar | | Philippines | | Thailand | |
	Male	Female	Male	Female	Male	Female	Male	Female	Male	Female
1950	22	11	32	18	17	12	13	25	22	11
1970	28	16	40	33	27	15	28	25	15	10
1990	44	37	44	40	31	21	40	37	28	27

Source: Xenos, Kabamalan and Westley 1999: 2–3; Xenos with Kabamalan 1998

was difficult to persuade parents to send their children to school or to instil in children who did attend school the importance of education. At that time the contribution that a child could make to the family work unit was deemed to have greater value than an education. By the mid-1990s, when Toyota conducted her fieldwork among the Akha, most parents were enthusiastic about the value of a modern education. Indeed, many adults, having missed out themselves, were busy taking evening classes (1999: 272).

This fundamental shift in the value attached to education is especially pronounced with respect to girls. Early efforts to promote education were almost solely geared to the education of boys. Today, at primary level at least, variations in provision between the sexes have narrowed considerably as enrolment rates among females have risen faster than among males (see Xenos and Kabamalan 1998a: 13, Xenos with Kabamalan 1998b). It is only at secondary and higher levels that a gender differential sometimes begins to reveal itself, and this too has narrowed considerably in the last few decades (Table 4.2). More important than gender in determining variations in levels of educational achievement today, is class (see, for example, Hefner 1990: 179–80). Where gender differentials do exist, they are often a product of income constraints rather than an ingrained belief that girls do not require or benefit from education beyond primary level.[8] It has also been noted that it is often eldest daughters who suffer most. They are withdrawn from schooling early to enable their younger siblings – and especially male siblings – the opportunity to continue past secondary level (Rodenburg 1997: 163 and see Tacoli 1996: 22).

The longitudinal Central Luzon Loop Survey shows a dramatic increase in the number of children in rice farming households in this part of the Philippines north of Manila entering college (higher education).[9] In 1986 just 7 per cent of working members of the sample farms had a college education; in 1990 the figure was 12 per cent; and by 1994, 20 per cent (Estudillo and Otsuka 1999: 502). The authors postulate that this 'remarkable increase' is due partly to increased family wealth and partly to the increased return to education from

[8] Of course, the fact that it is girls who are withdrawn from education when the economic burden becomes too great indicates, at the very least, that greater emphasis is placed on educating male offspring.
[9] For more background details on this survey see page 78.

non-farm jobs. This has encouraged parents to invest more heavily in their children's schooling and especially beyond secondary level.[10]

Scanning through accounts of contemporary rural life in Southeast Asia, it quickly becomes evident that rural people, almost without exception, accept the critical importance of education. It is universally acknowledged that the path to progress involves some measure of education – and the more, it seems, the better.[11] There is a keen thirst for education and parents are willing to forego most things to ensure their children acquire one. Families uproot and move to urban areas to take advantage of the better schools there; children are packed off to stay with relatives in distant towns; mothers sell their gold jewellery and fathers pawn or sell their land to raise the money for school fees; the economic value of a child is sacrificed; income is diverted from productive agricultural investments to paying for uniforms, school equipment, books and transport; and consumption expenditure is held back (see Rodenburg 1997: 151–2, Singhanetra-Renard 1999: 78, Fegan 1983: 39–40, Sherman 1990: 50–51, Wong 1987: 165).[12] Sometimes it can seem that this emphasis on education is single-minded to the detriment of wider considerations. Why are families willing to expend so much effort, invest so much money, and compromise their lives to such a degree, all in the name of education? There are clearly a number of considerations at work here.

To begin with, agriculture is perceived to be an occupation with little future. For many rural families, who operate only small landholdings, this is difficult to refute. Incomes from agriculture, already often meagre, have stagnated and it has only been by exploiting non-farm opportunities that many families have managed to meet their rising needs. As the popular expression in Northern Vietnam has it: 'it is not enough to buy a can of beer, even [when] you sell 10 kilograms of rice' (quoted in Tana 1996: 5). One study found that over half (52 per cent) of those working in the refuse and scrap recovery industry in Hanoi were doing so to supplement insufficient farm incomes (DiGregorio 1994: 108). Ritchie notes, with more than a touch of irony, that the only wealthy farmers in the Northern Thai village of Ban Lek are those who have sold their land (Ritchie 1996b: 131). In short, it seems, there is little to keep many young people in the village. Studies from agriculturally depressed areas like central Lombok in West Nusa Tenggara, Indonesia sometimes contend that taking up employment beyond the village is the 'only hope [that villagers have] of improving on their household economic welfare' (Mantra 1999: 65).

Secondly, agriculture is not only perceived to have little future in pecuniary terms, but it is also a low-status occupation. This, though, is partly a reflection of the ethos of education itself which, with its emphasis on modernity, has served to undermine the standing of agriculture vis à vis other occupations.

[10] The authors further recommend that to facilitate the shift in occupational structure from farm to non-farm, secondary and college education should be promoted (Estudillo and Otsuka 1999: 520).

[11] See, for example: De Koninck 1992: 163, Wong 1987: 164–5, Rodenburg 1997, and Sherman 1990.

[12] 'While ignorant of her son's branch of study, one elderly woman [in the Toba Batak area of Sumatra] told me that she had sold what little wealth she possessed in the form of gold a few years ago to finance her son's education. The details differ, but many village women, through their industry and their secret savings, make it possible for their children or siblings to continue their education' (Rodenburg 1997: 164).

Having received an education, especially if it has involved living in the city, few young people are able to adjust to the hard manual work of the fields, even if they should wish to.[13] Farming, in the Philippines, has become 'the work of simple barrio folk' (Kelly 1999a: 298). Moreover, in many cases others villagers would view it as a waste if educated people returned to manual work and did not make the most of their (expensively acquired) education (see Sherman 1990: 52).

Thirdly, education is perceived to provide access to the growing number of non-farm jobs that offer not just status but also greater stability of income and security of employment. This is especially true of government jobs (Oey-Gardiner 1991) but extends to factory and other forms of non-farm employment. When Zarina left school in 1986 at the age of 17 she took a job in a toy factory in Kalim, some 25 kilometres from her home village of Paya Keladi in Malaysia's Kedah province. Zarina's initial salary of M$143 per month was less than she could earn by working in the fields (De Koninck 1992: 163–4). But her move into factory work was not propelled just by a desire to be modern and escape from agriculture. There were also some sound financial reasons for the decision, notwithstanding the apparently unattractive salary. Zarina could only hope to find agricultural work for perhaps two to four months each year. Factory work offered her a steady, year-round income. It also provided the prospect that her salary would rise as she acquired additional skills. It is because of the decline of agriculture and the corresponding rise of non-agricultural work that parents are willing to earmark funds that formerly would have gone into various agricultural investments, for education. Thus Sherman, writing of the Batak in North Sumatra, observes that for villagers 'having children in school is perhaps the most promising means, for themselves or their descendants, to tap or connect to the power and wealth that vitalize the world beyond the border of the lake [Toba]' (1990: 50).

Fourthly, mothers and fathers are also, no doubt, influenced by the state pronouncements regarding the value and utility of education. Education permits people, in the state view of things, to make the transition from ignorance to light. Those villagers without education are, in Indonesia, often viewed as *bodoh*, or uneducated and ignorant; people with education, by contrast, are *maju*, or progressive (Vickers 1996: 30). It is education which drives this mental transformation.

The education of children is a valuable social marker for parents, bestowing a degree of social capital. Fegan (1983) notes from the Philippines that arranged marriage negotiations now take into account the investments that a family have made in the education of the child. This is seen as improving the life

---

[13] Fegan writes of students in the Philippines who, having graduated, 'would not stoop to field work while on "stand-by", waiting for a civil service post' (Fegan 1983: 40–41). Toyota quotes one young Akha woman who had left her village home for Chiang Mai: 'I never liked working in the fields; it is dirty and tedious. Although the wages are as low as an agricultural labourer, I much prefer to work as a shop assistant in town. I do not want my skin to get dark like my mother's from working outside. Dark skin is not beautiful. My mother's life is so pitiful [her husband is a drug addict and beats her]. I want my life to be better than hers' (1999: 303). See Bryceson (1997: 9) who makes much the same point with reference to Africa.

Plate 9 Woman power: taking vegetables to the daily street market in Rantepao, South Sulawesi, Indonesia.

chances of a young person. Therefore parents, for both social and economic reasons, often prefer to invest their scarce financial resources in education rather than in, say, farm equipment (see for example Leinbach and Del Casino 1998: 212). Finally, there is a strong moral imperative in many Southeast Asian societies for children – and especially daughters (see the discussion above) – to look after their parents. Investing in a child's education is akin to investing in a pension scheme. It ensures support in later years. Income generated by children working away from home is channelled back to the core rural household to a notable degree.[14] The other side of the coin from a child's obligation to their parents, is the parents' responsibility to the child. As one Northern Thai informant observed: '[i]n the past, the parents' prime responsibility and concern was to feed their children, today . . . [it] is to educate their children' (Bencha Yoddumnern-Attig 1992: 20).

As will have been clear from the last paragraphs, there is a circular element to this sequence of arguments: educating children takes them away from the fields. It imbues them with the belief that agriculture is beneath them. It provides them with skills which, because of the lack of employment opportunities for the best educated in many rural areas, serve to keep them away from home. Rodenburg (1997: 143–6) makes the point that in her study area in the Batak area of North Sumatra, while education is given extraordinary emphasis by the inhabitants of the area, there are virtually no jobs that require high levels of education. In other words, if the newly educated youth are to cash in their (and their families') educational investment they must leave their homes. Preston concurs with this view in his study of the Central Cordillera in the Philippines (1998: 377) while Kato, working in Negeri Sembilan, Malaysia argues that education, the New Economic Policy, and the availability of alternative occupations have, taken together, 'conspired to extinguish the young generation's interest in agriculture . . .' (1994: 167, see also Cederroth 1995: 173). Thus education, while it may be motivated by a belief that agriculture does not offer much of a future for the young, serves to further undermine agriculture. In some instances it seems that the insidious nature of what Rodenburg terms the 'education syndrome' has created a situation where any young people who have remained at home, in the fields, are automatically labelled 'failures' (Rodenburg 1997: 168–70).[15]

Returning to the generalisation at the beginning of this section, one consequence of this emphasis upon education is that some families are, apparently, willing to 'sacrifice' their land in order to achieve a higher educational status for their children. This, more than anything, illustrates the shift in priorities from farm to non-farm.

---

[14] There is some suggestion that this is coming under pressure as modernisation fragments families, socially and spatially. Singapore's Maintenance of Parents Bill, debated and passed by Parliament in 1994, is aimed at enshrining the duty of children to look after their parents in retirement (Rigg 1997: 144). But, significantly, parents can only call on the Bill's legal sanctions if they have been dutiful parents.

[15] Paulson and Rogers note this in their work in Western Samoa where they write that 'young Samoans who live away and return for visits (often bearing montary gifts) are accorded more respect and appreciation than those staying with the rural family to work in the family plantations' (1997: 183).

Plate 10   Local transport in Danang, Vietnam. Motorbikes, pickups and minibuses have transformed the ability of rural people to access opportunities beyond the local area.

Transportation, integration and migration: bridging the gap

In the preceding chapter the prevalent notion that before the modern period rural people in Southeast Asia were isolated from the market and the state was contested. It was suggested that the degree of interaction with the state and with the market have been consistently understated. This is not to say, however, that improvements in transport and communications have not had profound effects on rural areas and rural society.

The provision of rural transport services in the developing world has not been a topic that has received a great deal of attention (Johnston 1998: 1). And yet there has been a transport 'revolution' in Southeast Asia (see Dick and Forbes 1992, Singhanetra-Renard 1999: 72). There are few people in the region today who are isolated from the market or, for that matter, from the tendrils of the state. Even in countries like Laos, where the road system is still highly underdeveloped, access by people to the market, and vice versa, is proceeding rapidly (see Rigg 1997).

Some rural areas in Southeast Asia are not so clearly integrated into the mainstream, but nonetheless are at the cusp of intense cultural and economic change. This is evident in Healey's description of outlying villages on Aru, itself an outlying portion of marginal Maluku in Indonesia. Here, poor villagers, primarily engaged in subsistence-oriented cultivation and fishing, send their children to school, where the curriculum is the same as anywhere else in Indonesia. The essentials of modern consumer culture can be seen, heard and purchased in the town of Dobo. 'The social isolation of the self-sufficient local village community is thus overlain [by] its encapsulation within, and participation in, the nation state and wider modern world.' '"Modernity"', Healey continues, 'is very much ingrained in the prevailing conditions of the village' (Healey 1996: 23)

Improving transport and communications has helped spread the ideology of development (Booth 1995), creating a conduit along which ideas, beliefs and aspirations, along with money and goods, have been channelled. For Dearden, the provision of an efficient transport network has 'probably done more to change the landscape of the North [of Thailand] and the mindscape of [its] inhabitants than any other single factor' (Dearden 1995: 118). In Mae Sa, for example, the arrival of the mini-bus in the 1970s gave local women the oppor-tunity to market their agricultural produce in Chiang Mai city, 13 kilometres away. It also brought the city's better schools and expanding employment opportunities within commuting reach of the village (Singhanetra-Renard 1999: 72). As such, roads and cheap transport facilities transformed the economic (and other) opportunities open to Mae Sa's inhabitants. As several studies have shown, the distribution of poverty has an important spatial component. High levels of poverty in Java (Mason 1996) and Vietnam (van de Walle 1996) are closely linked to geographical marginality and levels of infrastructure provi-sion. Where transaction costs are high, it seems, the challenge of ameliorating poverty becomes that much more difficult.

Leaving home has long been noted as an important rite of passage for males, a means by which a young man can gain experience of the wider world, acquire a modicum of sophistication and, perhaps, accumulate some savings with which to build a house. Women, it has been said, were often reluctant to marry men who had stayed at home. What is significant is that the culture of modernity and a loosening of traditional constraints on female mobility have now not only permitted women to leave home, but created the social con-ditions where they are encouraged to do so. In Thailand, most rural migrants to Bangkok prior to the early 1980s were male. But by the middle of the 1980s surveys were showing that as many as two-thirds were female (Parnwell 1993: 8–9). The pattern of industrialisation (textiles, electronics) was preferentially selecting women workers over men while the easing of traditional constraints on female mobility was creating the social conditions where migration was sanctioned. Furthermore, 'livelihoods for . . . a large and increasing number of people . . . are becoming dependent upon the continued economic dynamism, and associated income-earning opportunities, which exist elsewhere [i.e. in urban areas, and in particular in Bangkok]' (Parnwell 1993: 9). In the Sumatran Lampungese settlement of Tiuh Baru girls who work in modern factories in Tangerang acquire beauty (pale skin, attractive clothes, jewellery), desirable knowledge (experience, confidence), and wealth (savings, household goods) by leaving home. They become, in the process, more attractive and valuable on the marriage market (Elmhirst 1998b). When factory migration from Tiuh Baru began in the early 1990s, parents were reluctant to let their daughters leave for the city, fearing that their moral status would be compromised in some way. By the late 1990s mothers and fathers, far from trying to dissuade their daughters, actively supported their engagement in factory work far from home.[16]

Not everyone is sanguine about the insidious infiltration of cultures of consumption into the countryside and the displacement of rural people from their villages. Malaysian popular singer Hang Mokhtar, in *Pulang Lah Orang Mudo* ('Go Home, Young People') laments the abandoned lands, the empty *surau* (Muslim prayer halls) and the villages devoid of youth and implores young people 'Better go home . . . Than go on migration without any clear purpose' (quoted in Kato 1994: 168–9). In the Philippines, it is common for the older generation to bemoan the emergence of the *barkada* or 'gang' as a forum in which youth subcultures can flourish. The *barkada* is a 'social context in which behaviour can stretch social norms – a crucible for redefining the aspirations and identities of youth, and a controlled rebellion against the overbearing institutions of family, lawfulness and hard work' (Kelly 1999b: 66). For the young, the *barkada* can be viewed as liberating, giving them the power to confront established and entrenched norms. It is used, for example, to explain how young women can work away from home against the wishes of their parents and make their own consumption decisions. For parents the emergence of the *barkada* explains the changes in attitude towards the church, sexual relationships, drinking and drug-taking, and work (Kelly 1999b: 66).

Work on extended metropolitan regions, or EMRs, in Asia has shown how in certain densely settled parts of the region, like the areas surrounding Manila, Jakarta, Bangkok, Ho Chi Minh City (Saigon) and Hanoi, improvements in transport and communications technology have led to a tight integration and interaction of people and activities.[17] The Asian component is seen to arise because of the juxtaposition of zones of high-density rice cultivation with the urban core. As scholars involved in this work have been at pains to point out, EMRs do not just designate areas where certain physical manifestations of the metropolitanisation of the countryside are in evidence, but also a process where the extreme fluidity and mobility of the population leads to intense interaction of commodities and people, and a tight articulation of agricultural and non-agricultural activities (Ginsburg 1991: 38, McGee 1991).

But, as I have argued elsewhere (Rigg 1997: 265–6), the emphasis on EMRs as physical entities and the preoccupation with the material, have tended to overlook the fact that in a functional sense there is no 'edge' to the EMR. Jamieson has suggested that the island of Java is on the verge of becoming one vast, extended metropolitan region (Jamieson 1991: 277). But even this seems to understate the degree to which the processes and effects of the interactions reach far beyond the physical boundaries of these regions. In Thailand, there are virtually no areas which are more than an 18-hour bus journey from the Bangkok EMR while in Indonesia's case the influence of the Jakarta-centred

[16] Elmhirst's work is discussed in more detail in Chapter 6.
[17] See McGee 1989 and 1991 and Ginsburg 1991 on EMRs in Asia in general; McGee and Greenberg 1992, Luxmon Wongsuphasawat 1995, and Parnwell and Luxmon Wongsuphasawat 1997 on Bankgok; Drakakis-Smith and Dixon 1997 on Ho Chi Minh City and Hanoi; and Jamieson 1991 on Java.

Box 4.1   Bangkhuad: from village to suburb, buffalo to Toyota

Howard Kaufman's study of Bangkhuad in the Central Plains of Thailand illustrates the importance of communication in changing villagers' access to work and the market. His initial research was undertaken in 1954 with a return period of fieldwork in 1969.

Even in 1954 there were signs that Bangkhuad would soon become functionally (and physically) engulfed by nearby Bangkok. Poorer villagers were already working in the city as truck drivers and labourers and he notes that the mobility permitted by improving roads and bus services was 'resulting in a gradual change from a sedentary socio-economic life to a migrating, extra-village, socio-economic life' (1977: 212). But he did not foresee the depth and pace of the changes that would be wrought over the next 15 years.

Outlining these momentous changes, Kaufman notes that 'The new and improved roads had served as catalysts for the rapid cultural upheaval in Bangkhuad, their greatest impact being felt in the economic sphere' (p. 219). Traffic began to come to the village more regularly, small factories were constructed within 'a short radius' of the village, which became a source of labour, land values soared and the sale of land escalated. To reach Bangkhuad's *wat* (monastery) in 1954 took a 30-minute car ride and then a 30-minute walk across paddy fields. By 1969 the journey could be made by car from the heart of Bangkok in 20 minutes. Interestingly, Kaufman also hints at the EMR hypothesis in his section on 'The new configuration of suburbanization' (pp. 230–1) when he talks of the Bangkhuad villager becoming 'unwittingly' part of Bangkok's suburban configuration. In summary, Kaufman writes that by 1969:

> The socio-economic perimeters [of Bangkhuad] had disintegrated, and the limits were now amorphous. The urban rural continuum, with its concept of gradual assimilation and metamorphosis, was not applicable. Within the short span of six years [between 1963 and 1969], Bangkhuad had leaped from the buffalo to the Toyota Corolla age and was enmeshed in the gears of modernization and 'progress' (1977: 232).

The author grieves for the loss of the old Bangkhuad. He is generally romantic about the loss of the old. As he writes: 'it [1971] was a time of paucity and a time of plenty, a time of large spendings and a time of large savings, a time of improved technological communications and a time of impaired family communication, a time of inceasing noise and increasing silence, a time of hope and alacrity, a time of despair and anxiety' (1977: 233).

EMR reaches far beyond Java. This is evident, for example, in Elmhirst's work on the settlement of Tiuh Baru in a relatively remote corner of the Sumatran province of Lampung, mentioned above:

> ... improved overland transport has led to the development of close links with West Java and export-oriented industrial areas around the edge of Jakarta, which may be reached in less than 18 hours overland and by ferry [from Tiuh Baru]. These links mean that people are able to circulate between areas, extending the influence of Indonesia's principal industrial zone into what has hitherto been a remote and inaccessible area (1998a: 4).

The introduction to this chapter outlined a link between infrastructural, cultural, and economic change. Improving communications, and particularly public road transport, are a vital background factor in the equation. Consumerism, education and the ideology of modernity provide key motivations for change. And the expansion of non-farm opportunities, both *in situ* and *ex situ*, provides the means of meeting these intensifying and expanding desires.

Leaf (1996) neatly combines these different elements of change in his account of BMW culture in the Jakarta extended metropolitan region and the blurring of the rural and urban (a theme which will be returned to more than once in this book). The 'revolusi Colt' brings vendors and small-scale traders into the city from the far periphery daily. The emergence of consumerism and the expansion of capitalism across the country under the influence of Suharto's New Order have replaced the traditional peasant culture of *cukupan* ('sufficiency') with 'expectations of steadily rising living standards' (p. 1622). The urban *kampong*, once an expression of rural society in the city, has become emblematic of the 'outward expansion of the urban economy'. For Leaf it is the intersection of these changes – in communication and transportation technologies, in desires, and in economies – which lies at the heart of the erosion of what, hitherto, has been the classic and entrenched dualism in Indonesian society and economy: between city and countryside.

Chapter 5

Stubbornly rural, tenaciously agricultural? Questioning the big picture

Introduction

In his book *Age of extremes: the short twentieth century 1914–1991*, Eric Hobsbawm contends that in the mid-1980s there were only three regions of the globe which 'remained essentially dominated by their villages and fields: sub-Saharan Africa, South and continental South-east Asia, and China'. He continues: 'In these regions alone was it still possible to find countries which the decline of the cultivators has apparently passed by . . .' (Hobsbawm 1994: 291). This chapter aims to show that at the beginning of the twenty-first century this perspective, so far as Southeast Asia is concerned, is increasingly difficult to sustain.[1] Southeast Asia is no longer a region of peasants. Indeed, the rest of this book will set out the case for a profound transformation in the nature and structure of the rural economy and rural lives. Angeles-Reyes, in her discussion of rural change in the Philippines, poses a question that sets the stage for the discussion in this and later chapters: 'What does the growth of nonfarm rural employment signify beyond the obvious fact of labour market differentiation?' (1994: 133). As will become clear, it is both symptomatic of deep-seated structural and social changes while also providing the impetus which further embeds such changes.

That rural areas of an economically fast-changing region should be undergoing deep transformation is not remarkable or surprising. As Table 5.1 shows, agriculture has seen its share of output fall dramatically over the last three decades. While a decline in the relative contribution of agriculture to gross domestic product will not be reflected in an equal change in the sectoral distribution of the labour force, we would expect to see a series of associated changes. These include:

- a shift of the labour force out of agricultural pursuits
- a loosening of the links between land and wealth

[1] Alexander *et al.* (1991), writing about Java, make the point that hitherto it has been accepted that non-farm activities were insignificant in rural livelihoods. However, they question the statistical base on which such an assumption is founded. The large residual categories in the data sets on rural employment and the relative invisibility of women's work have tended to shield, in their view, the true scale and significance of such work.

Table 5.1 Structure of the Southeast Asian Economies, 1970, 1980 and 1998 (% of GDP by sector)

	Agriculture			Industry			Services		
	1970	1980	1998	1970	1980	1998	1970	1980	1998
Brunei	–	–	–	–	–	–	–	–	–
Cambodia	–	–	51	–	–	15	–	–	34
Indonesia	45	24	16	19	42	43	36	34	41
Laos	–	–	52	–	–	21	–	–	27
Malaysia	29	22	12	25	38	48	46	40	40
Myanmar	–	47	59	–	13	10	–	41	31
Philippines	30	25	17	32	39	32	39	36	52
Singapore	2	1	0	30	38	35	68	61	65
Thailand	26	23	11	25	29	40	49	48	49
Vietnam	–	–	26	–	–	31	–	–	43

Sources: World Bank 1997, World Bank 1993, World Bank 2000

- an increase in the role played by labour earnings in determining household income
- an increase in the share of non-farm income in total income
- a movement of people from rural to urban areas
- changes in agricultural methods and cropping patterns.

This chapter will lay out the broad parameters of rural change in the Southeast Asian region, concentrating on economic indicators. Later chapters will provide the detailed exposition, examining how these broad transformations have impacted, for example, on household structure, decision-making, agricultural methods and labour allocation.

Questioning the statistical picture

An examination of statistical tables released by international agencies such as the World Bank and the United Nations Development Programme, or national bodies like the Thai National Statistical Office and Indonesia's Central Bureau of Statistics, reinforces the image that Southeast Asia remains a region where agriculture dominates the lives of most people and where, apparently, agricultural pursuits dominate the countryside (Table 5.2). Taking the simple average of the countries with a significant rural sector (i.e. excluding Brunei and Singapore), it seems that in 1997 68 per cent of the population were still living in rural areas and, in 1990, 61 per cent were engaged in agriculture. Scholars have noted the tendency both to view the region as 'quintessentially agrarian' (McVey 1992: 7) and to '"agrarianise" the countryside' (Alexander *et al.* 1991: 1). In this way, Southeast Asia becomes a region of farmers and farming, and the countryside becomes equated with agricultural pursuits. While acknowledging the deep structural changes that have occurred in the economies of the region, scholars

Table 5.2 Rural people, agricultural lives

	% of population living in rural areas			% of labour force in agriculture		
	World Bank	UNDP		World Bank	UNDP	
	1995	1970	1997	1990	1970	1990
Brunei	–	38	31	–	12	2
Cambodia	79	88	78	74	79	74
Indonesia	66	83	63	57	66	55
Laos	78	90	78	78	81	78
Malaysia	46	66	45	27	54	27
Myanmar	–	77	74	–	78	73
Philippines	47	67	44	45	58	46
Singapore	0	0	0	0	3	0
Thailand	80	87	79	64	80	64
Vietnam	79	82	81	72	77	71

Sources: World Bank 1997, UNDP 1998, UNDP 1999

at the same time appear reluctant to accept that this has also led to an equally deep transformation in rural people's lives.

Thus Angeles-Reyes, in her discussion of the Philippines, while remarking on high levels of rural-to-urban migration nonetheless maintains that the Philippine economy is 'predominantly' rural, employing 60 per cent of the population and generating a significant portion of export revenues and gross domestic product (1994: 134). Indeed she argues that 'many' parts of the country remain largely subsistence in character (1994: 135) and, on the basis of these observations, argues that there has been no fundamental structural transformation in the economy since 1960. Dixon, in his excellent review of the Thai economy, takes a similar tack to Angeles-Reyes when he writes that while Thailand might have achieved rapid economic growth during the 1990s 'observers appear to have lost sight of the fact that the country remains an essentially agricultural one . . . [and] . . . the agricultural sector . . . remains the principal source of livelihood for some 60 per cent of the population' (1999: 140).[2] The essence of these perspectives seems to be that while the countries concerned may have experienced economic change of some considerable magnitude, this has not undermined the essential agrarian character of the region.

But while country-level statistics released by national and international organisations may offer an image of a stubbornly rural and agricultural region, even in the face of significant economic change, local-level studies tend to offer

[2] But, in an earlier discussion in the same book, Dixon questions the accuracy of the country's employment statistics, observing that the official data may mask a 'much more substantial decline in the [agricultural] sector's importance as a source of livelihood' (1999: 19–20). This is discussed in more detail below. Nipon Poapongsakorn's analysis of change in the Thai rural labour market mirrors Dixon's perspective in acknowledging the growth of non-farm work and remarking on low rural incomes while at the same time offering the view that even in the dry season the agricultural sector accounts for 60 per cent of total employment (1994: 168).

a rather different image. Such studies appear to show that employment and residency patterns have, indeed, also undergone significant change. However, and for ease if nothing else, it is the country-level data that tend to be quoted in the literature and accepted by most commentators and agencies as indicative of the structure of employment and the geographical distribution of population. The argument offered here is that we should take greater note of the local studies and the stories that they tell about the nature and pace of change.

How has this misrepresentation or distortion of the true state of affairs, if indeed it is so, arisen? There seem to be five critical areas to consider in this connection. First, household registration systems in the region are inadequate given the high levels of mobility typical of many population groups. Second, labour force surveys are characteristically inaccurate. Third, definitions of rural and urban tend to be too static, failing to pick up the extension of urban forms into rural areas. Fourth, the categories used – rural/urban, agriculture/industry/ services – are insufficiently nuanced to take account of the nature of lives and livelihoods in modern Southeast Asia. And fifth, work has tended to ignore or play down the growing role of rural industries.

Household registration systems and labour force surveys

Household registration systems in the region significantly overestimate the proportion of the population living in rural areas and, by extension, underestimate the numbers living in urban areas. There are two core reasons for this state of affairs. First, the registration systems in place make the implicit assumption that people are relatively static and, that when they do move, they move on a permanent basis. And second, these systems are overly bureaucratic, inhibiting people from registering a change of residence once they have made a move (permanent, semi-permanent, or otherwise).

The dynamism of rural livelihoods, as people switch in and out of farming, is clear in labour force surveys conducted at different times of the year.[3] In Thailand the report of the labour force survey for 1998 shows that during the year the proportion of the total labour force employed in agriculture rose from a low of 39 per cent in May to a high of 51 per cent in August (Table 5.3).[4] This reflects the movement of people in and out of agriculture through the year in accordance with the changing seasons. May marks the end of the dry season when in most parts of Thailand rice cultivation is impossible, while August is in the middle of the wet season and the peak period for rice agriculture. Similar patterns of seasonal mobility are evident in Vietnam where, for example, 59 per cent of the peak season labour force in Hanoi's refuse recycling industry are temporary residents in the city – refugees from the countryside – whose arrival reflects patterns of labour demand in agriculture (DiGregorio 1994: 108) (Figure 5.1).

[3] This is a point that Fisher *et al.* (1997: 26–7) make with reference to India.
[4] Significantly, the annual labour force statistics for Thailand are estimated using data from round 3 of the labour force survey which is conducted in August – the peak agricultural period.

Table 5.3 Seasonal changes in the Thai labour force (1998)

	Feb	May	Aug	Nov
Labour force (million)				
Total	29.4	28.5	32.1	31.0
Agricultural	11.6	11.1	16.5	15.1
% Employed in				
Agriculture	40	39	51	49
Non-agriculture	60	61	49	51

Note: Feb = round 1, May = round 2, Aug = round 3, Nov = round 4
Source: assorted issues of the *Report of the labour force survey*, National Statistical Office, Bangkok

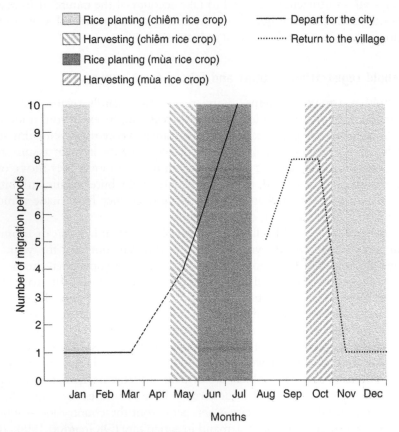

Figure 5.1 Mobility and the seasons, Vietnam
Source: Based on DiGregorio 1994

A linked difficulty concerns the accuracy of labour force surveys. Most such surveys are based on small samples. These are then scaled up and the results extrapolated from previous census data. Given the rate of economic change in the region it has been suggested that the results 'can, on occasion, be grossly inaccurate' (Young 1995: 653).[5] This has been highlighted by Dixon (1999: 19–20) as a problem in Thailand, where he believes the labour force in agriculture may be overestimated by as much as one-fifth. That this is an issue of a wider, generic nature is indicated by Young's work (1995) on Taiwan, South Korea, Singapore and Hong Kong and Rawski and Mead's on China (1998). In the latter case, the official figures 'overstate actual farm employment by a large and growing amount' with a 'margin of error [that] may exceed 100 million workers' (Rawski and Mead 1998: 767). These 'phantom farmers', as the authors style them, are the product of a system for estimating the size of the farm labour force from data on rural population, participation rates, and organised non-farm employment. In each instance the methods are dubious and the data inaccurate, with the result that official figures systematically overstate the size of the agricultural work force. Detailed studies consistently show in China (as in the countries of Southeast Asia), a marked decline in agricultural work and a pronounced movement of people off the land. Like Vietnam (see below), migrants in China are required both to obtain permission from their local administrative authorities before they take up work beyond the village and register their arrival in an urban centre. But studies show that only between a fifth and a half of the country's floating population do so (Roberts 1997: 269–70).

The bureaucratic deficiencies are perhaps starkest in the transitional economies that have inherited their registration systems from a command era when population movement was sharply restricted. Notwithstanding considerable economic reform, Vietnam continues to rely on its pre-reform household registration system. In Vietnam, moving to a new place of work or residency still requires the individual to register their move with the local People's Committee. But Tana, in her study of rural migrants to Hanoi found that only three out of 200 respondents asked their local authorities for permission to move and even fewer registered their arrival in Hanoi (1996: 31). An official survey by the Ministry of Labour of migrants to Hanoi and Haiphong found a higher level of registration, but even this recorded non-registration figures of 69 per cent and 73 per cent in sending and receiving areas respectively (Tana 1996: 31). Tana quotes from a 1993 statistical survey which found that '930,000 people in Vietnam either did not register or registered in one place but actually lived in another' (1996: 7). This figure included 350,000 in Ho Chi Minh City and 140,000 in Hanoi – and Tana believes that the true figures (the statistical survey was incomplete) are double even these estimates. It is widely accepted that the household registration system in Vietnam has broken down (Drakakis-Smith and Dixon 1997: 24) and yet central government continues to use the system as

---

[5] This is taken from Young's (1995) paper 'The tyranny of numbers' which begins, disarmingly, with the admission that this 'is a fairly boring and tedious paper . . .' (p. 641).

a key source for ascertaining patterns of change in employment and residency. Moreover, and as in other countries of the region, the Vietnamese census only measures long-term and permanent moves – and therefore fails to pick up the temporary and circular migration that appears to be the dominant form of mobility in the country (Anh Dang *et al.* 1997: 333)

A similar state of affairs exists in Thailand where workers who move to the city rarely register a change of residence. This is partly because they view their sojourn away from home as just that – a sojourn, and not a change of residence, even when they have been absent for five years or more (see page 91 for a fuller discussion). But equally important are the bureaucratic complexities needed to negotiate the country's notoriously rigid household registration system. It requires a visit to government offices in both the province of origin and of destination. And for those who are tenants – as many industrial workers are – there is the added requirement of ensuring their landlords enrol and support their application (Fairclough 1995: 17). Taken together, this means that several million urban workers are registered as living in rural areas. For example, a joint report by the Bangkok Metropolitan Authority and the Thai Development Research Institute published in 1991 calculated that two million migrants moved from rural areas to Bangkok each dry season (quoted in Paritta Chalermpow Koanantakool and Askew 1993: 28). The official population of the Bangkok metropolitan area in 1991 was 5.6 million, so this is clearly a substantial number (NSO 1993).[6] The same under-reporting is noted in studies from Indonesia where rural residents wishing to travel are officially obliged to obtain a letter of permission (or *surat keterangan bepergian*) from the village secretary. Yet in his study of the village of Slendro in Central Java Firman (1994) shows that there is little correspondance between recorded and actual levels of mobility.

Figures on temporary international labour migration are also highly unreliable. This applies right across the Asian region (Hugo 1993: 39–40). In 1995 Indonesia's Ministry of Labour recorded that there were 224,364 official Indonesian overseas contract workers. However various other sources put the number working in Peninsula Malaysia at 600,000, in Sabah and Sarawak (East Malaysia) at 350,000, with a further 40,000 working elsewhere in the region (Hugo 1997: 273). In other words, the true level of Indonesian overseas contract labouring is over four times greater than official figures suggest.[7] Most of these individuals, it seems, move from rural areas where their departure is not recorded.

---

[6] During the November 1996 elections in Thailand, the out-going government of Prime Minister Banharn Silpa-archa proposed a four-day public holiday so that Bangkok's army of migrant workers whose registration still recorded them as living in their rural villages could return 'home'. Commentators noted that Banharn's Chart Thai Party drew much of its support from the countryside and it was suggested that the holiday was required so that Banharn's natural constituency would have the opportunity to vote.

[7] In 1992/93 the Indonesian Ministry of Labour processed just under 30,000 application from workers moving to take up jobs in Malaysia and Singapore. Yet estimates of the number of Indonesians working in Malaysia ranged, at that time, from 200,000 to 700,000, and there were probably another 90,000 working in Singapore (Hugo 1993, see also Mantra 1999, Rigg 1997: 126–9).

Categorising lives

Employment statistics place people into categories of employment by sector. Thus it is normal for labour force statistics to list those employed in agriculture, industry and services. Sometimes different categories are conflated so that, for example, industry and services are presented as 'non-agriculture' and employment as a binary system: either you are a farmer or you are not. The evident deficiency of this approach is that it assumes that individuals are employed in a single sector when the accumulating evidence is that many 'farmers' are also engaged in non-farm work: 'Rural is contrasted with urban, agriculture is contrasted with industry, and people are interpreted as making complete transitions from one type of activity to another as development proceeds' (Ellis 1998: 1).

Household registration systems also tend to force people into single categories. In China the household registration system defines a household as either 'agricultural' or 'urban resident', with no room for flexibility (Roberts 1997: 254). Even when the status of an area is changed from rural to urban following city expansion existing residents are still stuck with their original registration.

Definitions of rural and urban

A second reason to doubt the data on the balance between rural and urban populations relates to the tendency for cities and towns to sprawl beyond their official boundaries. In this way many populations and settlements which are defined as rural might, in fact, be better classified as urban. Drakakis-Smith and Dixon (1997: 24) and McGee (1995: 256) both make this point with reference to Vietnam and it is well established in other countries of the region. Officially, the urban population in Vietnam grew by 1.4 million between 1989 and the end of 1994, implying a decline in the level of urbanisation from 19.9 per cent to 19.4 per cent – a highly unlikely scenario given the generally accepted direction of migration flows and pace of urban growth in the country (Drakakis-Smith and Dixon 1997: 24). Working in the Hanoi-Haiphong and Ho Chi Minh extended metropolitan areas of north and south Vietnam respectively, Drakakis-Smith and Dixon observe:

> Field work . . . together with discussions with officials, planners and consultants, suggests that urban land-uses and economic activities have moved extensively into the surrounding areas of the Red River Basin and Mekong Delta. Industrial, commercial and residential developments have spread along routeways, engulfing already densely populated rural areas, connecting and expanding large numbers of smaller centres (1997: 26).

The role of rural industries

While most industrial activity is located in or close to urban areas, rural industrialisation is significant and growing (see Chapter 8). Rural industrialisation is not limited to small-scale, village-based, craft-oriented industries but also includes modern, large-scale and often export-oriented industrial activities. Such industries offer rural residents the chance to work in industry by daily commuting from home.

Not only are industrial activities permeating the countryside, but urban residents and industrial workers are beginning to move out of the towns and cities and into rural areas. The countryside is becoming suburbanised. Singhanetra-Renard notes this in her work (1999) in Northern Thailand and so do Courtenay (1988) in Melaka and Maurer (1991) in Central Java. The village of Tirtonirmolo, located on the agriculturally rich Bantul plain, is being consumed by the fast-expanding city of Yogyakarta (Maurer 1991: 99–101). The village has become 'quasi-urban' in form and a 'dormitory village for white-collar workers and bureaucrats working in Yogyakarta' (1991: 101). While the number of household heads whose primary occupation is own account farming actually increased between 1972 and 1985 from 136 to 150, this farming rump was swamped by an influx of civil servants and other workers looking for an amenable place to live. Between 1972 and 1985 the number of household heads living in Tirtonirmolo and working as civil servants, commuting to their jobs in Yogya, increased from 93 to 314 (see Table 5.8 on p. 76).

The view from the village

As the discussion above has demonstrated, there are a number of reasons to question the picture provided by official statistics regarding the geographical and sectoral distribution of the population and labour force in Southeast Asia. To rectify this situation, we have to turn to village and local-level studies. It has been argued for some time that 'it is through detailed local-level studies that better conceptual understandings ... will develop' (Hart et al. 1989: 2). This focus on the local, though, raises another set of problems and difficulties and, in particular, the problem of selection. Over the years, thousands of studies of village economic and social change have been undertaken by assorted geographers, economists, anthropologists and sociologists. These studies tell multiple, often seemingly conflicting stories about the detailed processes and direction of change. For it to be otherwise would be remarkable. However, while in their detail the narratives are importantly different (and, to be sure, the devil is in the detail), they do appear consistently to show a shift in the balance of the village economy from farm to non-farm. In many cases this has occurred to the extent that the contribution of agriculture to household income is exceeded by non-farm income.[8] Thus, while Eder's warning that the 'safest (and most useful) conclusion for the moment ... is that the local dynamics of agrarian change and differentiation are immensely variable, often reflecting

---

[8] While this book is about rural change in Southeast Asia, studies from other countries of the developing world show similar trends in non-farm work. See, for example: Lanjouw (1999) on rural Ecuador (40 per cent of total income from non-farm work); Chandrasekhar (1993) on West Bengal (six villages with a range of 38–87 per cent of total income from non-farm sources); and Reardon (1997) on Africa (25 case studies with a simple average of 45 per cent of total income from non-farm sources).

Table 5.4 Longitudinal studies of rural change in Southeast Asia

Source	Location and years
Indonesia	
Maurer 1991*	Central Java, 1972–87
White and Wiradi 1989*	West, Central and East Java, 1971–81
Malaysia	
De Koninck 1992*	Kedah, 1981–87
Tsubouchi 1995*	Kelantan, 1971–91
Philippines	
Banzon-Bautista 1989*	Central Luzon, 1977–84
Eder 1993*	Palawan, 1970/72–88
Estudillo and Otsuka 1999*, Herdt 1987	Central Luzon, 1966–94
Hayami et al. 1998*; Hayami and Kikuchi 1981*	Laguna, 1974–95
Preston 1998*	Cordillera, Luzon, 1984–95
Thailand	
Rigg 1998	Northeastern Thailand, 1982–94
Singhanetra-Renard 1999	Northern Thailand, 1976–96
Tanabe 1994*, Ritchie 1993*, 1996a* 1996b*	Northern Thailand, 1974–91

* Work referred to in detail in the discussion

unique combinations of complex and even conflicting processes' (1993: 649) remains germane, that does not mean that we should shy away from the task of highlighting general themes and processes. But, having made a case for the utility – and feasibility – of picking out common threads, this is not to imply that it is possible to reduce all the studies noted below to some sort of 'average'. Difference and variation make it important to maintain the integrity of each study. This is particularly true, as will become clear in the next two chapters, when it comes to explaining the trends identified.

One point to note about these village studies *in toto* is that comparatively few are longitudinal. Ellis, in his survey article of rural livelihood diversification notes not only the general absence of work on the composition of rural household incomes but stresses, in particular, 'an almost total lack of data sets that are comparable across time intervals greater than two or three years' (1998: 8). While such studies are thinly represented for Southeast Asia – Ellis' paper, it should be noted, concentrates on sub-Saharan Africa – they are not totally lacking. The following discussion is based upon a selection of such longitudinal studies conducted in Indonesia, Malaysia, the Philippines and Thailand spanning different lengths of time but in each instance concluding between the mid-1980s and mid-1990s (Table 5.4).[9] Notably absent from the selection are studies from Cambodia, Laos, Myanmar and Vietnam. As it has only been comparatively recently that social scientists have been allowed back

[9] The exception to this is White and Wiradi's (1989) influential study which spans the years 1971–81.

into the field in these countries, longitudinal work is generally absent. A second point to note about the longitudinal studies discussed below is that they are not directly comparable. Not only do the studies focus on different countries (and regions of countries) and span different time periods, but they also adopt different techniques and measure different things. Most are village studies, but not all. Some take a simple sample of the population to measure change, while in other instances the original sample is followed up on successive occasions. Some assess changes in the contribution of different activities to household income, while others examine changes in the structure of employment. For some the emphasis is on agricultural change, while others focus on the non-farm side of people's livelihoods. These differences are noted, where appropriate.

Box 5.1   What is the rural non–farm sector?

In the literature there is general confusion about the terms used to describe different forms of rural employment and economic activity. These include: off-farm, non-farm, field, off-field, non-agricultural, own-farm and on-farm. There is considerable overlap between the terms and, sometimes, a deal of ambiguity (see the table). Furthermore, there is no internationally accepted definition. As Fisher *et al.* (1997: 8) note, there is a tendency to define things other than agriculture in negative terms – in other words, in terms of what they are not (non-agricultural, non-farm). Some studies make a distinction between landowning and non-landowning households, preferring to use non-farm to refer to the activities of landowning households only, and non-agricultural to refer to both landowning and non-landowning households. This distinction is avoided here because of the practical difficulty of clearly distinguishing between the two. The blurring of activities and the fluidity of households makes any clear-cut terminology workable only in theory.

Terms employed and (some possible) definitions

Off-farm	Work undertaken off one's own farm but which may be either agricultural or non-agricultural
Non-farm	All non-agricultural work, whether undertaken on-farm, in the local vicinity, or extra-locally
Field	All agricultural work, including processing activities, whether undertaken by a landowner on their own land or by a non-landowner
Off-field	Similar to non-farm, including all non-agricultural work, whether undertaken on-farm, in the local vicinity, or extra-locally
Non-agricultural	Corresponds closely with non-farm except that such work is undertaken by both farm and non-farm (i.e. those without any connection to the land) households
Own-farm	Agricultural work on a person's own landholding
On-farm	Non-farm work undertaken at 'home' – for example home-based piece-work

For discussion of the terminological morass see: Cederroth 1995: 104–7; Fisher *et al.* 1997.

Table 5.5 Occupational change in Paya Keladi and Matang Pinang, Kedah, Malaysia (1975–86)

	Paya Keladi			Matang Pinang		
	1975	1981	1986	1975	1981	1986
Sample population	163	156	158	144	157	144
Householders working on own farm	56	44	32	71	66	58
Householders earning agricultural wages	–	24	12	–	36	22
Householders engaged in cooperative work	–	8	0	–	26	9
Adult householders engaged in non-farm work	–	30	50*	–	8	11*
Former householders engaged in non-farm work	–	–	26*	–	–	4*

* These figures are for 1987
Source: De Koninck 1992

Breaking out of the village and escaping agriculture

De Koninck's work (1992) in two villages in Kedah on the Malay Peninsula undertaken on three occasions between 1975 and 1987 illustrates the trends that the national statistics tend to mask (Table 5.5). De Koninck's study is unusual in that for the second and third periods of fieldwork the original sample was maintained (i.e. the original households were tracked down and re-interviewed). In both Paya Keladi and Matang Pinang the survey results reveal a steady decline in the number of householders working their own land and wage labouring in agriculture. At the same time there is an increase, albeit modest, in the numbers of householders engaged in non-farm employment.[10] However, there is a contrast between the two villages in the degree to which householders have been able to gain access to non-farm jobs beyond the locality. Paya Keladi is strongly articulated with the wider region and the rapid decline in the contribution of agricultural employment (p. 185) is closely linked to the availability of local and non-local employment opportunities.[11] In Matang Pinang, by contrast, agriculture remained, at the time of the final survey in 1987, the main pillar supporting villagers' livelihoods. Even so, De Koninck anticipated that Matang Pinang's agricultural base was also being gradually undermined while he predicted that Paya Keladi would experience a further erosion in the importance of farming to villagers' livelihoods: 'The likelihood is . . . strong that an increasing number of the descendants of the peasants of Paya Keladi and Matang Pinang will need to break out of the community circle, if only to break out of agriculture, particularly in Matang Pinang' (1992: 186).[12]

[10] Occupations include: teachers, clerks, soldiers, factory workers, carpenters, merchants, general labourers, and traders (De Koninck 1992: 178).
[11] The superior educational status of residents in Paya Keladi is identified as important (p. 185). This links back to the discussion in Chapter 4.
[12] Although De Koninck has not returned to Paya Keladi and Matang Pinang since 1990 he suspects, from scraps of information gathered on agricultural decline in Kedah and Penang, that the trend identified in the 1992 study has continued (personal communication, 29.7.99).

Table 5.6 Primary occupations of household heads, Santa Lucia, Philippines (1977 and 1984)

	1977		1984	
	Number	%	Number	%
Farm	**149**	**47**	**162**	**35**
Rice farmer	82	26	114	25
Vegetable farmer	8	3	2	0
Farm labourer	59	18	46	10
Non-farm	**170**	**53**	**303**	**65**
Construction (home)	83	26	130	28
Construction (overseas)	17	5	78	17
Transport services	26	8	54	11
Trading	12	4	16	3
Employee*	26	8	18	3
Thresher (contractor)	1	0	2	0
Teacher	5	2	5	1
Total	**319**	**100**	**465**	**100**

* Includes: policeman, photographer, tailor, barber, printer, baker, textile worker, worker, poultry worker and piggery worker
Note: there are rounding errors in this table
Source: Banzon-Bautista 1989: 149

Another longitudinal study which, like De Koninck's, maintained the same sample households is Banzon-Bautista's work (1989) on Santa Lucia in Pampanga, Central Luzon in the Philippines and about 80 kilometres from Manila. Two surveys were conducted, in 1977 and 1984. Even in 1977 around half of the household heads in the sample reported that their primary occupation was something other than farming. By 1984 the figure had reached almost two-thirds (Table 5.6). Particularly dramatic was the increase in overseas contract labour migration to the Gulf where males from Santa Lucia were largely employed as carpenters in the construction industry (the Pampanga area has a particular reputation for its carpentry). In the case of Santa Lucia the primary force propelling agrarian change was not only embedded in the non-farm sector, it was also non-local and linked to what Banzon-Bautista refers to as the 'Saudi connection'.

In Tsubouchi's community study (1995) of Galok, a Malay village in Kelantan on the other side of the Peninsula from Kedah, where De Koninck undertook his fieldwork, the changes have been even more pronounced. The village has been visited on four occasions since 1970, most recently in 1991. During this two-decade period the number of households increased from 145 to 211, but the number engaged in the three core agricultural pursuits of rice cultivation, rubber tapping and tobacco cultivation decreased by between one-third and two-thirds (Table 5.7). Furthermore, the intensity of cultivation also declined. At the same time the number of villagers working in non-farm activities such

Table 5.7   The decline of agriculture, Galok, Kelantan, Malaysia (1971–91)

	1971	1984	1991
Households	145	157	211
Households engaged in:			
Rice cultivation	71	–	36
Rubber tapping	94	–	53
Tobacco cultivation	124	–	40

Source: data taken from Tsubouchi 1995

as the local saw mill or as teachers or migrant labourers in Singapore increased significantly.

Preston's work (1998) in a pair of villages on either side of the Cordillera (Central Highlands) of Luzon in the Philippines spans a ten-year period between 1984–85 and 1995. Poitan is in Ifugao province and Sagada in Mountain Province. The focus in this study is on the gradual decline and improverishment of agriculture as labour is preferentially allocated to other pursuits. Preston identifies a series of significant changes in agricultural methods and practice which indicate a shift in labour allocation out of agriculture. In Poitan, pondfields are no longer used for rice agriculture and in some cases have been abandoned, leading to a serious deterioration in their productive capacity. In both Poitan and Sagada, swidden cultivation of the hillsides has been effectively abandoned, while in Sagada cash cropping of vegetables has eclipsed rice cultivation for home consumption.[13] These changes have been brought about, in part at least, by a serious shortage of labour in the area as young people stay in education for longer, and men and women direct their energies elsewhere (i.e. in non-farm pursuits).

Maurer's (1991) analysis of change in four villages in the regency of Bantul close to Yogyakarta in Central Java spans the years 1972 to 1987. What is most striking in this particular study is the extent to which the researchers struggled to keep up with the pace of change that the farmers set for them. The original work in 1972–73 was set firmly within an agricultural modernisation framework and the intention was to measure and elucidate the social and economic impacts of the green revolution. During a succession of visits between 1976 and 1981 Maurer continued to enquire about agricultural modernisation-induced differentiation, only to find 'no real evidence of any strong land concentration or absolute impoverishment process' (1991: 97).[14] Rhetorically, he asked himself whether this could be explained by looking 'beyond the *sawah*'. The short answer was, 'yes'. Maurer notes a sharp decline in agricultural activities and a corresponding rise in non-farm occupations (Table 5.8). (In fact Table 5.8 underestimates the degree of change in the four villages – a point that the author

---

[13] Preston identified a similar series of changes in his work in Java (Preston 1989).

[14] The absence of any absolute process of polarisation resonates with Eder's (1993) and Hayami and Kikuchi's (1981) work also noted in this chapter.

Table 5.8   Structure of employment in four Central Java villages, Indonesia (1972 and 1985)

	Tirto		Timbul		Wukir		Argo	
	1972	1985	1972	1985	1972	1985	1972	1985
Total households	**828**	**985**	**616**	**675**	**667**	**682**	**434**	**409**
Landowning farmers	136	150	281	256	372	268	310	204
Tenants/sharecroppers	7	0	52	38	26	19	45	7
Agricultural labourers	48	7	98	104	24	35	37	87
Total farm	**191**	**157**	**431**	**398**	**422**	**322**	**392**	**298**
Industrial wage labourers	174	181	7	28	0	18	0	9
Artisans/daily workers	243	294	80	116	155	219	17	60
Civil servants	93	314	62	66	66	72	16	26
Traders/shopkeepers	127	39	36	67	24	51	9	16
Total non-farm	**637**	**828**	**185**	**277**	**245**	**360**	**42**	**111**
Main occupation (%)								
Farm	23	16	70	59	63	47	90	73
Non-farm	77	84	30	41	37	53	10	27

Note: occupations are listed as the main activity of the household head
Source: Maurer 1991: 98

makes – because it only records the primary occupation of the household head; it does not reveal the degree to which other household members are exploiting non-farm work.) While there are important differences between the four villages, Maurer identifies a common trend towards increasing reliance on non-farm work. Importantly, Maurer notes a contrast between the impact of non-farm employment on agriculture in environmentally rich and poor villages. Tirtonirmolo and Timbulharjo are situated in an area highly favourable for rice intensification. Wukisari and Agrodadi, by contrast, occupy marginal and poorly irrigated lands. In the former two villages a middle class peasantry of professional, commercial farmers has emerged, while in the latter it has not (1991: 107–8).

One influential study – in terms of understanding agrarian change in Southeast Asia at least – is White and Wiradi's (1989) analysis of nine villages scattered across West, Central and East Java spanning the years between 1971 and 1981. While the authors emphasise the differences between the nine villages distributed over much of Java from the far west to the far east, 'some broad generalizations', they suggest, 'are possible' (1989: 291). In particular, they note the dependence on non-farm activities for almost two-thirds of income. Only one village, where agriculture contributed two-thirds of total income, in their view, could be regarded as agricultural (1989: 294).

Rural industrialisation and diversification

Much of the literature on rural change has portrayed rural people being sucked out of their home villages, whether temporarily or permanently, to engage

Table 5.9 Percentage composition of household income in the village of 'East Laguna', Philippines (1974–95)

	1966	1974	1976	1987	1995
Number of households	66	95	109	156	242
Farmers	46	54	55	53	51
Landless	20	41	54	103	191
Farm		77	64	37	15
Rice		63	49	27	7
Non-rice		14	15	10	8
Non-farm		5	6	11	23
Commerce		4	–	7	6
Manufacturing		1	–	1	13
Transport		0	–	2	4
Wage labouring		17	30	45	43
Farm work		14	17	24	15
Casual non-farm work		1	⎰	7	13
Salaried non-farm work		2	⎱ 13	14	15
Other		1	–	7	19
Total		100	100	100	100
Farm*		91	81	61	30
Non-farm		9	19	39	70

* Includes rice, non-farm and farm work income
Sources: Hayami *et al.* 1998: 139, Hayami and Kikuchi 1981: 102

in work beyond the locality. However, there are also a significant number of cases where income from non-farm work has been earned *in situ* (the role of rural industrialisation is discussed in detail in Chapter 8).

Hayami and Kikuchi and their collaborators have been surveying the *barangay* (village) of East Laguna, in the province of Laguna, since 1974. The sixth of these surveys was undertaken in 1995 (Hayami *et al.* 1998). Laguna is one of the wealthiest and most productive agricultural regions in the Philippines and it was here that the Green Revolution first took hold. While the area shows a high level of landlessness, it has been portrayed as resolutely rural in character. Hayami *et al.* remark that, 'typical of rural villages in the Philippines, this village had traditionally been characterized by low reliance on nonfarm economic activities' (1998: 138). As this statement hints, the situation has changed. In 1974, 93 per cent of income in East Laguna came from agricultural endeavours. By 1995 this figure – apparently to the researchers' 'great surprise' – had declined to 30 per cent (Table 5.9). The authors explain this shift from farm to non-farm as resulting from the opening of a number of manufacturing establishments in the local area between 1991 and 1995, including a paper mill and several metal craft businesses. Furthermore, they identify a proliferation of similar manufacturing enterprises in the wider area since the late 1980s. Significantly, these non-farm opportunities were particularly exploited by landless households in the study area.

The village of East Laguna in Hayami *et al.*'s (1998) study is the same settlement that featured in Hayami and Kikuchi's *Asian village economy at the crossroads* (1981).[15] Looking back at the 1981 study we read of a village where rice farming is 'by far' the most important source of income and where non-farm opportunities are few (pages 101–3). Furthermore, East Laguna village, while it showed signs of growing inequality did not exhibit any tendency towards polarisation into a landless proletariat with a handful of large, commercial farmers.[16] The authors put this down to the basic institutional environment where 'traditional moral principles of mutual help and income sharing' have been preserved (p. 123). The pressure of people on the land in the area is illustrated by the rising proportion of landless households. In 1966 there were more than twice as many farming households as landless households in the village – 46 versus 20 (1981: 105). By 1976 the number of landless households exceeded the number of landed households, and by 1995 there were nearly four times as many landless as farming households (Table 5.9).

A rather different approach to assessing change over time is the 'Central Luzon Loop Survey', conducted at approximately four-year intervals since 1966 by the International Rice Research Institute (Estudillo and Otsuka 1999).[17] Central Luzon, like Laguna, is highly productive in agricultural terms and is often termed the 'rice bowl' of the Philippines. The survey selected approximately 100 rice farming households in a 'loop' through five provinces north of Manila. Significantly these were agricultural households – rural households not engaged in rice farming were excluded from the survey. With this in mind, the results are even more striking. In 1966, agriculture was of the 'utmost' importance to people's livelihoods, and rice farming formed the major component of this (Table 5.10). Non-farm income made up a little over a quarter of total household income. By 1994 this non-farm share had nearly doubled to 51 per cent. Furthermore, the shift from farm to non-farm accelerated during the 1990s – and this in a country which had the least dynamic (at that time) non-farm sector among the Asean members with a significant rural component. The authors conclude by stressing the structural shift in income among farming households from land to labour (i.e. from farm to non-farm). In the case of the villages surveyed the impetus behind this shift is put down to the development of the urban labour market and the increased access of farm households to that market. Furthermore, they note that if the demand for non-farm employment is to be met without exacerbating the urban problems already evident in Manila and other major Southeast Asian cities, then policies to promote rural industrialisation will need to be considered.

[15] The settlement is called 'East Laguna Village' in the 1981 study but given no name in the 1998 paper. Here East Laguna Village is used.

[16] Eder makes the same point in his study of San Jose on the Philippine island of Palawan (1993: 657).

[17] Although information on non-farm income was not always included in the survey – hence the pattern of years recorded in Table 5.10.

Table 5.10   Percentage composition of household income in Central Luzon, Philippines (1966–94)

	1966	1986	1990	1994
Sample size	92	120	108	100
Farm	**73**	**62**	**59**	**49**
Rice	57	46	38	39
Non-rice	16	17	21	10
Non-farm	**27**	**38**	**41**	**51**

Source: Estudillo and Otsuka 1999: 509

Preston's Cordillera study (1998) noted above draws an important contrast between the two villages of Poitan and Sagada in terms of the availability of local opportunities for occupational diversification. Poitan is within easy reach of the local centre of Banaue and, partly as a result, has seen a significant expansion in wood carving and weaving activities, keeping people on the farm but out of agriculture. In the case of Poitan an interplay of often self-reinforcing factors is at work: education and aspirational changes are encouraging young people (especially) to reject farming in favour of other occupations; lack of labour in agriculture is leading to a decline in the productive base; a decline in the productive base is reducing returns from farming; while shrinking farm incomes are encouraging people to look to non-farm pursuits to meet the shortfall.

While Preston presents a picture of agriculture under pressure in Poitan and Sagada, Eder's work (1993) in San Jose on the island of Palawan, also in the Philippines, argues for a complementary relationship between farm and non-farm. Despite far-reaching economic and demographic change between the two periods of fieldwork in 1970–72 and 1988 (including an increase in population from 112 to 278 households), the family farm was remarkably persistent – indeed it thrived (1993: 650). The period spanned by the fieldwork saw a proliferation of new economic opportunities in the area, including market vending, charcoal making, tricycle driving, wage labouring in the Palawan capital of Puerto Princesa City, and overseas contract employment. By 1988, 60 per cent of San Jose's farming households generated a significant share of their income through off-farm, non-agricultural activities.[18] But while in Poitan and Sagada such non-farm work undermined farming, in San Jose such activities dove-tailed neatly with agriculture, permitting small farmers to improve their standards of living even while their landholdings were diminishing in size. In the light of this, Eder argues that 'economic development in the region created

[18] A difficulty with Eder's work is that he defines 'agricultural income' as only income coming from work on a farmer's own holding. Thus agricultural wage labour on another farm is classified as non-agricultural. This tends to disguise the extent to which the balance between farm and non-farm income sources is shifting.

Table 5.11   Occupational change in Ban Lek, Northern Thailand (1974–91)

Occupation	1974		1985		1991	
	HH	%	HH	%	HH	%
Farming	113	48.5	127	44.1	5	4.8
Farming and wage labour	8	3.4	34	11.8	19	18.3
Farming and other	13	5.6	21	7.3	11	10.6
Wage labour	72	30.9	72	25.0	46	44.2
Self employed/entrepreneur	13	5.6	12	4.2	10	9.6
Salaried*	6	2.6	4	1.4	4	3.9
Other	8	3.4	18	6.2	9	8.6

* mostly in government employment
Sources: adapted from Ritchie 1993: 10 and 1996b: 125

a variety of opportunities for off-farm wage and self-employment, some of which opportunities were relatively renumerative and combined readily with agriculture' (1993: 651).

The experience of Ban Lek, in Northern Thailand's Chiang Mai valley is another case where occupational diversification has been based on a proliferation of non-farm opportunities within commuting distance of the village. This village has been studied since the mid-1970s when the Japanese scholar Shigeharu Tanabe worked there.[19] It is interesting to read his account (1994) of the village at that time, and his expectations of its economic future. In 1974 Ban Lek was a 'typical' agricultural community with almost 90 per cent of households defined as 'farming'. It was, he says, a village where economic activities were 'highly concentrated on agriculture' (1994: 107) and which was 'remote from the centre' (p. 251). But while Tanabe's book paints a picture of a 'traditional' Thai farming community he also provides some fleeting glimpses of the changes that were just over the horizon. Even in the mid-1970s only a minority of households (30 per cent) were involved in farming full-time and most were also engaged in a variety of supplementary occupations from trading and retailing through to handicraft production and wage labouring. These were concentrated, however, in the vicinity of the village and only a handful of people worked away from the village, in Bangkok, during the dry season. By the time Ritchie began his work in Ban Lek in the early 1990s it had become a very different place, one where the 'interpenetration' of farm and non-farm and local and non-local had deeply transformed the village's economic and social structures (Ritchie 1993, 1996a and 1996b) (Table 5.11). Radically improved communications and a burgeoning in employment opportunities in Chiang Mai City had led the 'supplementary' non-farm work of the mid-1970s to become the central component of household livelihoods. By the 1990s, in the 'typical' Northern Thai agricultural community of Ban Lek, only a few households could be easily classified as 'farming' – although farming might

[19]   In his work the village is called 'Chiang Mai village'; in Ritchie's work it goes by the name Ban Lek ('Small Village'). The latter name will be used here.

have touched most people's lives. As Ritchie writes: 'Although the village is in a rural setting, the people and households who are involved in agriculture are in the minority' (1996b: 123–6).

Indochina and Myanmar

It was noted earlier that there is a general absence of local level longitudinal studies of rural change from Cambodia, Laos, Myanmar and Vietnam. That said, there are studies which, sometimes through retrospectively constructing past conditions, identify similar processes to those outlined above. Tana (1996) and DiGregorio (1994) note the degree to which workers in specific industries and occupations such as construction, kitchen work, domestic service, cyclo-driving and scavenging are circular migrants from rural areas. In Tana's sample of 200 migrant workers in Hanoi, 70 per cent were paddy farmers while another 16 per cent combined rice farming with an *in situ* sideline occupation (1996: 24). Dang Phong's (1995) survey of five Red River Delta provinces in 1993 also identifies a diversification of activity and a general move out of rice agriculture. The reform process, he argues, has opened up far more opportunities for farmers beyond rice agriculture. But this has not led to rising urbanisation (he writes of *dérizaculturation de l'économie*) but, in his view, to *dérizaculturation de la population rurale* or an *in situ* diversification of livelihoods. Further, he believes this process can only intensify.

Conclusion

The clear lesson that comes through in the local studies summarised above is that the balance between farm and non-farm incomes has profoundly shifted over the last two to three decades. From the early 1990s, studies have *consistently* shown that non-farm incomes equal or exceed farm incomes in their contribution to the total household budget.[20] Reflecting this, there has been an associated change in the pattern of work, from farm to non-farm. Clearly, there is more to this than just a shift in the relative contributions of farm and non-farm activities and in the structure of employment for it to be labelled 'profound'. These deeper implications will be discussed in the following chapters. But even in the absence of this further discussion it is worth remarking, once more, on the contrast between the picture that is presented in the national statistics and the image that comes through in local studies. It is this which, perhaps more than anything else, explains the tendency for scholars to suggest on the one hand that Southeast Asia remains essentially agrarian in character while, at the same time, commenting on the deep changes that are occurring in rural lives and livelihoods.

Such a schizophrenic approach to interpreting rural change is seen in the conclusion to Elson's excellent book (1997) on the modern history of the peasantry

---

[20] In his survey article on livelihood diversification in sub-Saharan Africa, Ellis states that 'there seems to be an informed consensus that diversity has been increasing in recent history' (1998: 5).

in Southeast Asia. On the one hand he notes the deep changes that are occurring in rural lives. He also remarks upon the reluctance for people to admit that the Age of the Peasantry in the region has come to an end. But at the same time as writing that one era is at an end, and another beckons, he admits:

> It is true, of course, that Southeast Asia's populations are still overwhelmingly rural in residence and that the majority of the region's workforces find employment in spheres closely associated with rural production. It is also true that many Southeast Asians still construct their sense of self and community in ways not wildly dissimilar to that of their forebears a century ago (Elson 1997: 239).

The importance of these changes is that they hint at a profound change in the basis of rural differentiation. Hitherto, the tendency has been to see rural differentiation as an outcome of essentially agrarian processes. So, for example, a great deal of attention has been focused on the impact of agricultural commercialisation and, in particular, the effects of the dissemination of the technology of the Green Revolution on rural economies and societies. But what if agriculture is becoming, as the above studies appear to indicate, a subsidiary activity for many rural households? Where, then, are the roots of rural differentiation? Such studies indicate that the focus of work on agrarian change should endeavour to bring non-farm activities, which for so long have been marginalised at the edges as 'subsidiary' activities defined in negative terms, to centre stage.

Chapter 6

Hybrid households, fragmenting homes

Introduction

The previous chapter addressed the issue of the structural transformation of the countryside, using evidence from local-level studies to make the case for a profound change in the basis of rural life and livelihood. Studies from across the region show – consistently – an increase in the contribution of non-farm incomes to total incomes. What the chapter did not address was the question of how these changes have influenced (and are influenced by) the operation of the household, and the social processes and structures that constitute the household. Moreover, little attention was paid to the differential effects of these changes on different classes or groups in rural society – men and women, young and old, rich and poor. The discussion that follows, then, is designed to add flesh to the (largely) economic skeleton pieced together in the previous chapter. While diversification may be occurring on a broad (and global) scale, interpretations of what such diversification means are diverse and often conflicting. In a survey article concentrating on studies from sub-Saharan Africa, but which is equally relevant to Southeast Asia, Ellis warns:

> Diversification may occur both as a deliberate household strategy, or as an involuntary response to crisis. It is found both to diminish and to accentuate rural inequality. It can act both as a safety valve for the rural poor and as a means of accumulation for the rural rich. It can benefit farm investment and productivity or impoverish agriculture by withdrawing critical resources (Ellis 1998: 2).

(Re)considering the household

In Chapter 3, it was suggested that research since the early 1980s has questioned the notion of the 'village' as the basic building block of Southeast Asian society (see page 29). Essentially, this has taken two lines. First, scholars have challenged the antiquity of the village; and second they have asserted that the traditional village was less harmonious and more differentiated than previously imagined. As with the village, so with the household.

It remains common for studies of rural change not only to use the household as the basic survey component but to treat it as a single, welfare-maximising decision-making unit – to encompass, in other words, a single utility function

Table 6.1 Problematising the household

Deficiencies in traditional household analyses

- The household and the individual are merged analytically.
- Household strategies are represented as reflecting the common good.
- Decision-making is characterised as collective.
- The interests of the household and individuals are viewed as intersecting.
- Household heads are presented as altruistic.
- Little account is taken of cultural and geographical variations in the social construction of households.

(see Becker 1991: 230). As Russell says, 'it is household labour on household fields for household consumption' (1993: 756). This is most clearly associated with Becker's New Household Economics which has been characterised as assuming that 'each household pursues one overarching collective goal that reflects a common set of interests' (Wolf 1992: 14). Furthermore, 'the household decision maker is portrayed as a benevolent dictator . . . who has internalized family members' needs, makes decisions with the collective good in mind, and rules with justice' (Wolf 1992: 15). Certainly, the neoclassical tone that permeates Becker's work, together with its emphasis on optimisation, rationality, economism and joint utility, does appear to justify such a characterisation. But a close reading of his *A treatise on the family* (1991) makes it clear that Becker does allow for rather greater intra-household exploitation and conflict than summaries of his work indicate.[1] 'Malfeasance within a family', he writes 'is not simply a theoretical possibility but one that has been recognised for thousands of years . . .' (1991: 48).[2]

Nonetheless, since the 1980s this view of the operation of the household as a single, welfare-maximising decision-making unit has come under close scrutiny for a variety of reasons. These include: spatial fragmentation, heightened tensions and conflicts between genders and generations, implicit assumptions of permanence and predictability, and a lack of attention to cultural difference (see Wolf 1992: 12–20, Hart 1992 and 1995, Folbre 1984 and 1986, Li 1996, Chant 1998: 8–9, MacPhail 1992, Koopman 1991, Radcliffe 1986, Redclift and Whatmore 1990) (Table 6.1). In the light of these deficiencies, some scholars have argued that the concentration on the household as the basic social and economic building block has obscured the true nature of rural society and its organisation, and that its use as a unit of analysis should be more cautious and critical – and possibly even avoided (see Russell 1993).

---

[1] There are parallels here with the moral economy/rational economy debate discussed in Chapter 3 (see page 31).

[2] In reply to Ester Boserup's criticisms of his work he argues that 'an efficient division of labor is perfectly consistent with exploitation of women by husbands and parents – a "patrimony" system – that reduces their well-being and their command of their lives' (1991: 4).

While the household remains, in Southeast Asia at least, the key social unit, it is now widely accepted that it should not be viewed as undifferentiated and, certainly, not as unproblematic.[3] The household is composed of individuals whose interests are not only different, but may be at odds. The result is tension and conflict. This, though, is not sufficient reason to reject the household as a useful unit of analysis. Most Southeast Asians consider themselves to be members of households, and the household, as a collective enterprise, is far from dead. Rather the point is that the character of the household has become – and probably always was – rather different from that which standard economic analyses have traditionally assigned to it.

It is not possible to discuss households and individuals in households interchangeably (Wolf 1990: 45–46) or, as Folbre puts it, to treat the household as an 'individual by another name' (1986: 20), because of the conflicts and inequalities which are such a part of its operation. As I have written elsewhere: 'the household remains a stage where cooperation and conflict, corporatism and individualism, mutuality and inequality, and consensus and discordance, co-exist. . . . the household is defined by dissonance' (Rigg, forthcoming[a]). Rather than rejecting the household, scholars have problematised it.[4] Indeed Wolf, in her work (1992) on women and migration in Java, notes that the household fills a useful intermediary position between analyses which stress the individual, and ignore the place and role of the individual in wider social and economic structures, and unremitting structural analyses which consign the individual – and the household – to virtual invisibility (1992: 13, see also Wolf 1990 and Rodenburg 1997). As Chant writes in her paper on women, the household and migration 'even in instances where women ostensibly make their own decisions to migrate, it is hard to abstract household conditions from the process . . .' (1998: 12). Households, she continues, 'not only [create] the material conditions for gender-selective migration but also [act] as filters for familial gender ideologies which impact upon motives for migration and the relative autonomy of migrant decision-making' (1998: 12).

Difficulties arise when the importance of the household in understanding patterns of action is taken one step further to imply that there are 'household strategies' in which the interests of different members of the household intersect for the 'common good'. From this position it is another small step to treating the household as an undifferentiated unit of production, consumption and reproduction. As the following discussion will show, the interests of household members do not coincide. 'Strategies' are contested, negotiated, imposed and subverted. What is to the advantage of one member may undermine the position of another. Moreover, there is a good deal of fluidity as changing circumstances create the space for strategies to metamorphose. But the recognition that the household is a contested space does not undermine the position of the household as the primary unit of analysis. The trick, as Preston says, 'is

---

[3] As recently as 1990 Diane Wolf found she could write: '. . . Third World households appear to be the only context in which the myth of family solidarity and unity is perpetuated, and this is seen most clearly in the context of household strategies' (1990: 43–44).

[4] Wolf talks of 'cracking open the "black box" of the household' (Wolf 1992: 22).

to collect information that, in some way, recognizes individual and collective scales of action in households' (Preston 1994: 207)

Gender and generation

The most extensive body of work directed at reconceptualising the household has been undertaken by scholars interested in the gender dimensions of household relations. In many cases these discussions also overlap with generational concerns. Moreover, it is not just a case of conceptual advances demanding a new look at the operation of the household; development has accentuated and amplified existing divisions and brought new pressures to bear on the 'traditional' household. Factory work, for example, confers on young women a higher status, not to mention greater spending power and financial autonomy, and also endows women with greater assertiveness and independence (see Wolf 1992: 192–3). Intra-household economic differentiation, spatial dislocation, and shifting gender and generational relations are all serving to accentuate the shortcomings that were inherent in the unitary approach to the household that characterised traditional work.

Scholars like Wolf (1990, 1992) in Java, Hart (1994) in Malaysia, and Elmhirst (1995, 1996, 1998a, 1988b) in Sumatra have all highlighted, in their different ways, an important shift in intra-household power relations. In each case unmarried daughters from rural communities were found to have significant and growing decision-making autonomy, particularly with regard to employment decisions. In Wolf's case, she found that daughters were often willing to defy their parents' wishes not to take up factory employment and 'dutiful daughters were the exception rather than the rule' (1990: 50). While girls aged 13 or 14 would comply with their parents' wishes, older daughters 'usually disobeyed and rebelled' when it came to decisions concerning their labour (1990: 51). But autonomy was selectively granted. When it came to questions of sexuality or sexual reputation, for example, it was usual for daughters to toe the parental line (Wolf 1992: 194). On the basis of her work in the Muda area of Malaysia, Hart identifies a similar trend among young unmarried women towards greater autonomy and independence. Furthermore, she takes this one step further to argue for a 'reconceptualisation' of the household that takes into account the shift 'in the structure and exercise of power between men and women and between elders and juniors . . .' (1994: 49).

Elmhirst's work, which spans a period during which migration to factory work by young, unmarried Lampungese women in a rural community became established and then institutionalised, adds an interesting temporal dimension to the discussion. The migration of young women to Tangerang (an industrial area on the fringes of Jakarta) began in 1990. By 1994 around one-third of all girls aged 17–20 were going to the area, 'usually on their own initiative and after pressing their parents into agreeing to let them go' (Elmhirst 1998a: 11). So, over the space of just four years there occurred an apparently profound change in the position of unmarried daughters in a community where, traditionally, they had little autonomy and their mobility was sharply constrained.

There also occurred an important – and parallel – transition in factory work from being discouraged and exceptional, to accepted and commonplace. Elmhirst highlights a complex series of economic and social changes that permitted, condoned and then encouraged such migration.

Initially, 'young women were drawn towards work in the factories for a variety of reasons, few of which were purely economic, and none of which indicated their desire to help their families' (1998a: 11). The first few migrants were, undoubtedly, brave. But as the numbers increased so 'bored' young women were increasingly willing – and able – to push back the boundaries set by their parents. Like Wolf, Elmhirst notes how some simply took matters into their own hands. This not only had ramifications for the household concerned but also challenged the wider gender structures of the community and the views held about factory work and women's autonomy (1998a: 12, 1998b: 8). By 1998 factory work has become an important rite of passage for young women in Tiuh Baru – just as migration is viewed as such for males in many Southeast Asian communities. Furthermore, in the mid-1990s, girls were under little obligation to remit money to their families. Daughters, as Elmhirst puts it, were 'assumed out' of the household economic equation (1998b). But by the late 1990s parents expected daughters to contribute to the household budget and while some portion of earnings was earmarked for the daughters, a significant proportion was allocated to household expenses and the support of siblings' educational and other needs. The economic crisis dating from 1997 further strengthened parents' hold on their daughters' labour, providing a moral imperative to channel funds back 'home':

> ... daughters who were once seen as having no economic role ... are now [1998] regarded as 'unemployed' when they are not engaged in factory work. In 1994, the idea that daughters' role in the household should constitute 'unemployment' was unthinkable (Elmhirst 1998b).

In each of these cases, the availability of non-farm work for women plays a central role in explaining and permitting the changes outlined above. This, in turn, links into the political economy of industrialisation in countries like Malaysia and Indonesia, and thus into wider processes of globalisation. Many areas of labour-intensive export-oriented manufacturing favour the employment of women, and especially young women (see Rigg 1997, Lok 1993, Wolf 1992, Akin Rabibhadana 1993, Chant and McIlwaine 1995a and 1995b). A combination of physiological, psychological, socio-cultural and situational factors explains why firms often prefer to hire women rather than men. However, the inordinate attention paid to export-oriented manufacturing has, perhaps, led to an erroneous impression that women are displacing men from many areas of non-farm work. For the moment at least, it seems that there is considerable occupational segregation by gender. While women may be preferred to men in 'feminine' jobs in the textile, garment, footwear and semiconductor industries, 'masculine' jobs in many areas of industrial production remain dominated by men (Floro and Schaefer 1998).

The availability of work, and the additional income that it can generate for cash-strapped rural households caught up in the modernisation maelstrom with all its attendant consumption pressures (see Chapter 4), creates the economic context within which young women can break out of the household moral 'envelope' and countenance leaving home. Having left home and taken up such work, unmarried daughters acquire status and further assertiveness. In this way, the household represents a stage where local and global processes of economic change intersect with cultural practices and social norms.

Rich(er) and poor(er) households

There are, quite clearly, important changes under way in the structure and operation of the (rural) household in Southeast Asia. However, it is important to stress that these changes are being unequally experienced, and not just between countries and cultures. Village studies show that different households are adapting to changing circumstances in markedly different ways.

Koppel and Hawkins (1994) identify two broad trajectories of change in rural Asia. The first is essentially progressive and emendatory. In this developmental sequence, rising rural incomes, savings and educational levels stimulate demand for non-farm goods and services. Mechanisation of agricultural production frees farm labour for work in non-farm activities where higher wages stimulate further occupational mobility which, in turn, engenders additional positive changes in rural economies and livelihoods.[5] Their second trajectory is one of degeneration and debasement. In this sequence, stagnant agriculture and uneven access to land displace poor households from rural areas, forcing them into low-paid, low-status non-farm work. The process is propelled by poverty and lack of opportunity in agriculture. The first trajectory is associated with the experiences of Japan, Korea and Taiwan and the second with that of the countries of South Asia. In sketching out these two trajectories, Koppel and Hawkins are referring, in broad terms, to national trends. However, their identification of fundamentally different processes driving the shift into non-farm work also resonates with conclusions drawn from local-level studies in many parts of the world.

While polarisation in rural areas of Southeast Asia has not been quite as marked as expected, there has been a divergence in the income, and occupational characteristics of rich(er) and poor(er) households. The prevalence of households with non-viable, or sub-livelihood landholding is evident in numerous studies. In his work in the Luzon village of Manggahan, 110 kilometres south of Manila, Muijzenberg notes that three-quarters of all households receive little or no income from rice cultivation (1991: 330). The same is true of Firman's (1994) account of the Central Java village of Slendro, where he notes the reality that for *most* households having access to work outside agriculture and beyond the village is a matter of survival. This has been termed 'distress diversification'.

[5] Whether there is such a virtuous cycle of rural–urban interactions is discussed in more detail in Chapter 8.

But while studies may note the emergence of households with non-viable agricultural resources, this is not to suggest that all such households are in a similar predicament. Hart, for example, in the village of Sungai Gajah in Malaysia's province of Kedah notes that the group of households with non-viable holdings comprises both the poor and an emergent middle class (1992). Hart dates the rise of the middle class in Sungai Gajah to the late 1970s and links this with the diversification of livelihoods into non-farm work (in this case mainly for men).[6] In these middle-class families women are fully domest-icated, while for poor households with non-viable farms women take over agriculture and play a central economic role in supporting the family. This process of 'housewifisation' has also been noted by Kato (1994) and De Koninck (1992) in Negeri Sembilan and Kedah respectively, where they identify the emergence of the *seri rumah*, or 'princess of the house'.[7] Women, relieved of the need to work in the fields due to mechanisation and the proliferation of relatively well-paid non-farm jobs, have retreated to the house.

Divergence of household responses to livelihood pressures is also clear in Elmhirst's work in Sumatra, where she contrasts the experiences of a Javanese transmigrant community and an indigenous Lampungese village (the latter noted above) (Elmhirst 1995). In the Javanese village, farming alone was not sufficient to meet household needs and the shortfall was met by men and women taking up work in the local area, mainly in the logging and plantation industries. There were few constraints on female mobility and activity and women and men embraced a wide range of non-farm jobs in a flexible and accommodating fashion. In the Lampungese settlement, as noted above, few women work locally as such work is viewed as shameful and degrading. Instead, young women have reworked local *adat* (custom) by engaging in extra-local work in Tangerang. As one villager explained to Elmhirst, 'there is no shame in what cannot be seen'. For the Javanese transmigrants, work was a means to meet basic needs – a means of survival. For the Lampungese, at least until the Asian crisis of 1997, non-farm work by young unmarried women was not regarded as important for household survival and the income generated was generally discounted (again, see above).

The attractions of non-farm work vary according to the wealth status of each household. White and Wiradi, on the basis of their survey of nine villages in East, Central and West Java undertaken in 1981, draw a distinction between the livelihood strategies of rich, middle and poor households. In their view, rich households embraced non-farm work as a means of accumulating further wealth, reflected in a positive relationship between farm and non-farm

[6] Hart associates Sungai Gajah's new middle class with state patronage. Most of the better-paid jobs are in government service of one kind or another and access to these jobs is usually through well-connected village landowners who can access the resources of the state, particularly through strategic links with the United Malays National Organisation (UMNO) (1992: 819–20).
[7] There is also a strong state-directed project to create a social context where the nuclear family, with the woman as mother and housewife, is the accepted norm. In Indonesia this is known as the culture of 'ibuism' – housewifisation. Ibuism embeds women firmly in the domestic sphere, identifying their roles as faithful companion, household manager, producer of children (preferably two), mother and educator, and good citizen (Guinness, 1994: 283).

incomes. For middle-income households it was a means of consolidation, while for the poor, involvement in non-farm work was often driven by the necessity to survive (White and Wiradi 1989). Among these landless and land-poor households there was a negative relationship between agricultural and non-farm incomes.

While reducing households to just one of three income/wealth categories and then allotting to each a particular livelihood strategy harbours the risk of excessive reductionism, it does highlight the important point that individuals and households engage in non-farm work for different reasons. White and Wiradi's perspective has been endorsed by more recent work. Leinbach and Watkins in their analysis of Cinta Karya in South Sumatra, for example, propose that for some households *merantau* (circular migration) can be viewed as a means of maintaining basic survival while for others it is a strategy aimed at engendering financial advancement (1998: 50). Cederroth makes a similar point about patterns of non-farm work in Bantur, East Java where for the poor such work is described as a 'safety valve' and for the better-off as a means of propelling them from a traditional lifestyle to a modern lifestyle (1995: 113). Reflecting these different motivations, the form that non-farm work takes, and the returns to such work, vary enormously. For the poor, work may be highly marginal – generating no significant surplus (see page 131).

The conclusion that the role of non-farm work varies according to class (broadly defined) has been highlighted in a range of work from across the developing world. Bryceson, for example, in her discussion of de-agrarianisation in sub-Saharan Africa makes a three-fold division – mirroring White and Wiradi's work – of 'profit maximisation', 'income stabilisation' and 'risk minimisation' (1997a: 8, see also Bryceson 1996). That non-farm work may be propelled by very different forces, and therefore take very different forms, is also reflected in Smit's work on South Africa (1998), Francis' on Kenya (1998), Berry's on sub-Saharan Africa (1993), Zoomers and Kleinpenning on central Paraguay (1996), Krüger's (1998) on Botswana, Chandrasekhar's on West Bengal (1993), Fisher *et al.*'s on India in general (1997), and Saith's (1991) on the wider Asian experience. Some broader surveys of the contribution of non-farm work to total household income appear to reveal interesting variations across the developing world. Reardon in his analysis of 18 studies from Africa concludes, for example, that 'compared to poorer households, upper income strata households have much higher shares [around double] of nonfarm income in total income' (1997: 737). This, he notes, is the opposite of the situation revealed in work in South Asia. But it would be premature, on the basis of information to hand, to do more than flag this differential. For Southeast Asia at least the available evidence precludes any clear statement on the income-equity effects of non-farm work (see Leones and Feldman 1998: 791). Moreover, the permutations are becoming more complex as rural weath is becoming increasingly de-linked from issues of land ownership. Hart (1992) notes this for Malaysia and so does Ritchie (1996b) for Northern Thailand. The class of the rural rich includes both those households with large landholdings and those with small landholdings.

Spatial fragmentation

Not only has the household traditionally been viewed as undifferentiated; it has also been regarded as spatially situated and coherent – reflected in the use of the term 'co-residential dwelling unit'. Many national and international data sources continue to employ this spatially grounded perspective where the household is one of shared residence (Chant 1998: 7). Yet the processes outlined in the previous chapter and discussed in rather more detail above make it clear that the social differentiation of the household has been accompanied by increasing spatial fragmentation. But this does not imply that people are also becoming, in the process, socially disembedded. Household members living away from home, sometimes for extended lengths of time and at some distance, remain a component part of the household. It is usual for such individuals to remit money. They will often return, sometimes for key social events (weddings, funerals or important annual festivals) and sometimes for peak periods in the agricultural cycle. Many hold the intention of returning home, of marrying and settling down. More to the point, most do not view their *de facto* place of residence as 'home'. In this way, for functional economic reasons (contributions to household finances and labour), for social reasons (meeting religious and other cultural commitments) and for psycho-social reasons (notions of 'home'), absent members of the household are absent in name only. The degree of spatial dislocation is probably greatest in Thailand, and particularly in Thailand's poor and environmentally marginal Northeastern region where there is a long tradition of migrating for work.[8] In his study of 97 villages across the province of Yasothon, Funahashi writes that 'in most villages . . . older people and young children are conspicious, while it is rare to see young people' (1996: 108). In writing this, Funahashi is echoing the observations of numerous scholars, journalists and commentators.[9] Usually such migration is associated with the movement of young, unmarried daughters and sons. But even when a young, married couple leaves to live and work beyond the village, the ties with the natal village remain strong. It is common for the wife's sisters and parents to cultivate the young couple's land while they are absent and for the couple to send money home – a system that is known as *baeng kan pai haa ngaan*, or 'making money separately'.[10] Furthermore, should the couple have children it is usual for them to be cared for by the grandparents in the village (*liang lan*). This is something Akin notes in his study of Ban Talat Phra in Khon Kaen province, also in the Northeast of Thailand, where the 'entire middle generation has gone to work in Bangkok' and where 'the daughters bring their babies back for their parents to look after' (Akin Rabibhadana 1993: 20). When the maternal grandparents are too old to work the land the daughter and her husband will

---

[8] Most of these movements are short-term and circular and are not picked up in census returns.
[9] See, for example Akin Rabibhadana 1993: 20, Sanitsuda Ekachai's essay 'Silence in the village' (1990) and Singhanetra-Renard 1999: 83.
[10] A more common term than the one quoted by Funahashi is *yaek yaai kan pai haa ngaan*, which translates literally as 'separating out to go and look for work'. In other words, each person going off to look for work in different places (personal communication from Rachel Harrison).

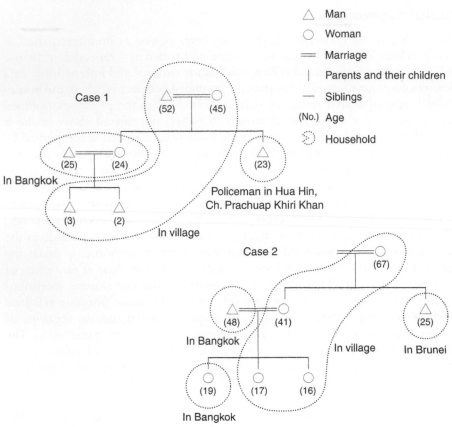

Figure 6.1 Extended households in Yasothon, Northeast Thailand

Source: Funahashi 1996

return home to take over the demands of farming and to take on the responsibility of caring for the aged parents.[11] Funahashi found that 11 per cent of households surveyed in Yasothon consisted of grandparents and their grandchildren and a further 6 per cent of older couples with their daughter and grandchildren. Indeed, the predominant household forms in the villages surveyed were older couples, older couples with grandchildren, and older couples with daughter and grandchildren (see Figure 6.1).

While Thailand may offer some of the most striking examples of what has been termed the 'geriatrification' of rural areas (see Redclift and Whatmore 1990: 185), the same trends are evident elsewhere in the region – although with important differences. Writing on the 'brawn drain' from the Muda area of Malaysia, Wong writes that 'Young men are drawn as individuals out of their households, rather than the household as such out of the village' (Wong

---

[11] Northeastern Thailand is traditionally matrifocal, although the system is in general decline. It was usual for the youngest daughter and her husband to take on the role of caring for the maternal parents. They would also, traditionally, have inherited the family house and land.

1987: 164 (she places all the sentence in italics)). This drain of young men from the village, in terms of scale, has been 'astounding'. By 1980 37 young men had left the village and only 22 unmarried young men remained. Moreover, the drain was not confined to a single social group. In Wong's case the selective migration of men can be linked to strong cultural controls on the migration of women – although work conducted since 1980 has, once again, shown how culture is both malleable and evolving.

The presence of social networks has been highlighted to explain the flow of migrants from often (apparently) remote rural communities to job opportunities, usually in urban areas and sometimes abroad. Such networks facilitate movement even when government policies are specifically intended to restrict citywards migration, and they also explain the discrete manner in which villagers from one locality (often community) are channelled into particular jobs and factories (see Hugo 1997: 280–82 and Rigg 1997: 212–14). 'Inter-personal networks founded on geography, ethnicity, kinship, friendship, membership, patronage and clientage mould the human landscape' (Rigg 1997: 212).[12] From the perspective of the discussion here, these social networks also, however, illustrate how tightly migrants remain tied to their villages of origin.

'Home' and the extended household

If households are spatially fragmented but socially embedded, where (and what) is home? Asking tuk-tuk, cyclo and becak drivers in Thailand, Vietnam and Indonesia where they live and where is home may be the prerogative of the lazy academic, but the answers are instructive. Home cannot be equated with where these people live. Assigning identity to place misses the key characteristic of many people's lives in the region: one of movement, and one of split identities and occupations, of lives lived in multiple ways – by job, sector, location, and age.

Similar difficulties arise when it comes to assigning an occupation to an individual. In 1959 in the village of Ban Ping in Northern Thailand to be a villager was to be a farmer – to 'make fields' (*het naa*) – an 'unconsidered, automatic, primordial' identity (Moerman and Miller 1989: 308). Today most of the inhabitants of Ban Ping, and many other villages, are no longer farmers in the pure sense, if at all, but the inertia of identity means that they may refer to themselves as such. In Ban Dong in Western Thailand's province of Uthai

---

[12] There is a great deal of literature on the role of social networks. See: Ulack (1983) on migrants in squatter communities in Mindanao, the Philippines; Paritta Chalermpow Koanantakool and Askew (1993) on slum residents in Bangkok, Thailand; Hiebert (1993) and DiGregorio (1994) on waste scavengers in Hanoi, Vietnam; Firman (1991) on workers on construction sites in Bandung, Indonesia; Ong on factory workers in West Malaysia (Ong 1987: 153–5); Wolf (1992) and Mather (1983) on factory workers in Java; Nipon Poapongsakorn on labour recruitment in agriculture in Thailand (1994: 184); Cederroth (1995) on finding a job in an East Javanese village; Hugo (1993) on international labour migration between Indonesia and Malaysia; Aphichat Chamratrithirong *et al.* (1995) on migration in Thailand in general; and Rigg (1989) on international migrants from Northeast Thailand. From beyond the region see Fall (1998) on migrants' long-distance relationships in Dakar.

Thani, for example, villagers describe themselves as *chao naa* – rice farmers – despite the fact that non-rice production in many cases is more important in sustaining livelihoods (Hirsch 1989a).

The most interesting work which addresses the issue of the relationship between identity, 'home' and mobility comes from the Pacific. In their study of migration among the Wosera Abelam of Papua New Guinea, Curry and Koczberski (1998) argue that 'a migrant never really leaves the village in a spiritual and cultural sense, and hence there is no clear cut distinction between sources and destination sites when examining the lived experience of Wosera migration' (1998: 36). They tentatively suggest that the destination site is an extension of home (p. 47). This resonates with Chapman's work. He quotes John Waiko, a Melanesian, who explains: 'We Melanesians are all engaged in a tapestry of life, where the threads of movement hold everything together' (1995: 257). Mobility, for some Pacific islanders, is so central to their lives that to equate place of residence with home is to ignore this defining characteristic.

To suggest that rural migrants remain strongly attached to their rural place of origin is also to argue that they are disengaged from their urban place of residence. This is reflected in the tendency for migrants to bring the country-side to the city and to cluster together in particular geographical areas and in particular types of occupation (see the discussion in Chapter 8). The presence of dormitories and boarding and lodging houses for rural migrants further strengthens the sense of rural identity, and the sense of separation from the urban mainstream (see DiGregorio 1994: 100, Sjahrir 1993, Wolf 1992). Tana, in her study of Hanoi, notes the dependence of rural migrants on fellow villagers for support – both social and financial – and also the lack of desire to 'sink roots in the city'. She argues that this indicates a 'blurring of distinctions between living at home and living in Hanoi' (1996: 45) and quotes Hanoians, remarking that the influx of rural migrants means the city is not urbanising (*thanh thi hoa*) but ruralising (*nong thon hoa*) (Tana, 1996: 45).[13] In his study of the scavenging and recycling industry in Hanoi, DiGregorio (1994) develops this further. He found that not only were workers in this industry sourced from particular villages and districts, but over 90 per cent of his sample lived in boarding houses, the majority of these in a single ward. For him, the 'commun-ity' is not geographical at all (in the spatial sense) but divided between rural and urban components. He hazards:

> **Boarding houses link urban and rural segments of the community. They provide a gathering point for village-mates, and a source of shelter, information, access to markets, and security. . . . Nearly all boarding houses in the community are segregated by village of origin, a situation that provides an extension of village relationships to the urban context. . . . Boarding houses also provide a point for the transfer of news and information between the rural and urban components of the community (1994: 101–2).**

[13] See Krüger's discussion of Botswana where he writes: 'When these strong mental links to the countryside persist for so many years, then parts of a traditional, rural value system must have been brought to the city and kept there . . .' (1998: 120).

It is factory dormitories which provide the most striking cases of migrant detachment from the mainstream of urban life (see for instance Firman 1991, Sjahrir 1993). Sjahrir quotes a report from the Indonesian language magazine *Prisma* on the operation of an *asrama*, or dormitory for female workers in a textile factory in West Java. After they have completed their shift, female workers return to their dormitory where '[t]hey are not allowed to meet or to communicate with outsiders' and are 'practically isolated from the outside world . . . [T]hey are there only for work . . .' (1993: 252). These young women, collectively a critical component in Indonesia's industrialisation efforts are, in a sense, outsiders at the centre. Physically a part of the city, structurally attached to the modern, industrial economy but economically, psychologically and socially connected to their rural roots.[11] It is this sort of tension between space and identity which has led scholars, for many years, to write of the 'ruralisation' of the city.[15]

While most people from the countryside, wherever they happen to live, may still view their natal villages as 'home', this is clearly not set in stone. It should be seen as a point on a historical continuum that is forever evolving. 'For some villages and former villagers', Moerman and Miller suggest, 'Ban Ping [in Northern Thailand] has become less a home than a place to leave behind, seek temporary refuge in, or send remittances to' (1989: 317). This is something that Macpherson also notes in his work among West Samoans residing in New Zealand. While, he suggests, they 'display considerable public affection for their homeland', and also reinforce this with remittances and donations of time and energy to the activities of their families and ancestral villages this should not, in his opinion, be taken to indicate that they will return 'home' at some later date (Macpherson 1985). Furthermore, over time migrants' commitment to their ancestral communities tends to decline as they become socially embedded in their new homes. It is difficult for migrants to invest time, energy and money in both places and there must come a time when they have to decide where their primary interests lie (Macpherson 1994).

Occupational multiplicity and the fungible household

Scholars of rural Southeast Asia have long recognised the importance of diversity in maintaining livelihoods in marginal and capricious environments. Multi-stranded livelihoods were the norm in areas like Northeastern Thailand and the Indonesian islands of East Nusa Tenggara and Madura. In Northeast Thailand farmers traditionally embraced a complex strategy of rice cultivation in which different types of rice would be planted on different types of riceland at different times of year (Grandstaff 1988 and 1992: 138, see also Rigg 1986). This was complemented by other activities from the planting of dryland crops

[14] In his work on Luzon in the Philippines, Kelly writes: 'Workers are temporarily abstracted from the locality but continue to provide economic inputs to local households . . .' (Kelly 1999b: 63).
[15] See Krüger on Botswana and the African city where he argues that 'cities in sub-Saharan Africa have *always* been somewhat ruralized . . .' (1998: 122 (emphasis in original)).

on higher land not suited to rice cultivation to the collection of non-timber forest products from surrounding wildlands. Grandstaff writes of the need for farmers to embrace a 'diverse portfolio of activities' in a resource-poor region like the Northeast. Fox's work on the island of Roti in Eastern Indonesia also emphasises the importance of flexibility and variety in maintaining subsistence when, in any one year, a part of the subsistence system might fail (Fox 1977: 50), while Dick argues that the rural population of the resource-deficient island of Madura have traditionally resorted to migration to top up their income and improve their livelihoods (Dick 1993: 14). For the poor, the need to engage in a number of activities to maintain subsistence was – and is – often even greater.

While traditional means of preserving diversity have been eroded as modern crop varieties have displaced local varieties and mono-crops are planted in preference to mixed assemblages of cultivars, the burgeoning of non-farm work has added a new avenue for promoting diversity. It is this new form of diversity, rather than the more traditional approach described by Grandstaff and Fox, that scholars like White (1976) and Muijzenberg (1991) are referring to when they write of 'occupational multiplicity', and 'diversification for survival'. Indeed, Hart goes so far as to propose that the 'perpetuation of multiple, diversified, spatially extended livelihood strategies . . . are a defining feature of late twentieth century capitalism' (1996: 269).

In the traditional fungible household resources and labour are efficiently and swiftly transferred between activities according to circumstances and priorities. However, the availability of extra-local non-farm work has accentuated the spatial component of fungibility, allowing households to re-allocate resources through space as well as through time and within the family (see Leinbach and Del Casino 1998: 196). Studies from many parts of the developing world, including Southeast Asia, show that the contribution to the household budget of household members living away from home may be greater than that of those who remain at home.

Leinbach and Watkins (1998) have developed this in the context of South Sumatra, where they liken remittance behaviour to investment portfolio theory. In their view, migration decisions are made by the family with a view to maintaining stability of income through embracing a diverse mix of activities that cut across genders, generations and space. Thus remittances become 'not a casual product of migration but an explicit and integral part of a family's economic survival strategies'. Fall (1998: 142) makes much the same point with reference to diversification behaviour among rural households in Senegal, which he views as being founded on a 'collective strategy' to diversify incomes sources and thus spread risks in a context where agriculture is in deep crisis.

This notion of the fungible household efficiently and rationally allocating resources according to need is in danger of falling foul of many of the difficulties noted above in the discussion of the household as a single decision-making unit. Individuals increasingly contest their allotted role, whether they are young women pressing to work in factories away from home or young men, endowed with superior education, resisting attempts to force them into farming. Furthermore, while occupational diversity may exist at the level of the household, opportunities for diversification are not equally open to men and women. Leinbach and Del Casino in their work in South Sumatra (1998) remark on the continuing gendered division of labour in Indonesia that permits men to engage in certain types of work (and patterns of work – see below), which are not open to women. While developmental pressures are eroding some of the barriers to female participation in the labour force, age and gender continue to play an important role in explaining patterns of work.[16] It is because of just such mobility-constraining reproductive demands that Hart makes a case (albeit with particular reference to South Africa) for spatial clustering of non-agricultural activities in rural regions, arguing that such spatial proximity cuts travel costs and 'makes possible the combination and sequencing of multiple activities' (1996: 270).

Studies have noted that diversification may involve households straddling the subsistence–commodity divide with some members remaining embedded in subsistence production while others engage in commodity production, whether in the local area or further afield. This perspective on diversification is particularly prominent in work on the Pacific Islands (see, for example, Paulson and Rogers 1997: 175, Curry and Koczberski 1998: 46, Curtain 1981). The desire to maintain a foothold in subsistence production is usually seen as a risk-minimising and/or stability-enhancing strategy. Paulson and Rogers, in their work on Western Samoa for example, contend that in their two study villages ('which are not atypical') most people maintain a subsistence base that has shown resilience and stability in the face of major disturbances (1997: 177). Keeping one's subsistence option open provides an exit option or safety valve when households dabble in the market and commodity production.[17]

It was widely felt that the Asian economic crisis would lead rural households to retreat into agriculture and, more particularly, into subsistence agriculture (see Rigg forthcoming(b)). Articles appeared in the regional press with titles like 'The countryside will save the day' (Chang Noi 1997) and 'Farmers come to the rescue' (Bullard et al. 1998: fn 73). The Forum of the Poor, for example, an umbrella organisation representing Thai farmers' groups, saw 'the crisis as an opportunity to go "back to the village" and slow the seemingly irreversible pattern of urbanisation and industrialisation, bridging the gap between urban and rural populations and re-establishing the traditional values associated with village and agricultural life' (Bullard et al. 1998: 22). This probably did not occur to the extent imagined in more populist circles (see the discussion in Rigg forthcoming(b)). Indeed, it can be argued that non-farm work effectively subsidises subsistence farming which would otherwise wither and die.

The family life cycle

The family life cycle is another important factor determining the ability of individuals and households to engage in non-farm work, and especially in

---

[16] See the discussion of the gender division of labour in agriculture in Chapter 7.
[17] Curtain, in his description of the 'straddled peasant household' in Papua New Guinea refers to a system of 'dual dependence' (1981: 199).

extra-local non-farm work. Households with young, dependent children of pre-school age are often highly restricted in terms of their mobility.[18] And even those with children of school age have to mould work around the demands of childcare. This can have a determining impact on livelihoods, especially in the case of land-poor households. It is for this reason that some studies prefer to label young poor households the 'striving' poor (ActionAid Vietnam 1995) or the 'aspiring' poor or 'pre-prosperous' (Cohen 1996). This tacitly acknowledges the dynamism that underlies a household's or individual's income status at any point in time (see Edmundson 1994 and Rigg 1998b).[19] Young households are often not only restricted in terms of their mobility, but may also be waiting to inherit land from elderly parents.

In Makarti Jaya, a transmigration settlement in South Sumatra, Ibu Sugiran is a trader with school-age children. When her children are at school she is able to travel much further to conduct business compared to when they are at home, at which times she works from her family compound. Because men are not expected to take on the responsibility for household management and childcare, restrictions on their mobility are far less pronounced (Leinbach and Del Casino 1998: 214). In this way, migration often becomes gender-selective, but particularly at later stages in the life cycle. Adult married women, and especially those with dependent children, exhibit a lower propensity to migrate compared with husbands, sons and unmarried daughters.[20]

This view is confirmed in another study of a transmigrant community in South Sumatra, where only two variables were significant in explaining the propensity to engage in *merantau*, or circular migration: age of household head and age of spouse (Leinbach and Watkins 1998: 49) (Table 6.2). Those involved in *merantau* in Cinta Karya were at least ten years older than those who were not, and also had older children. Furthermore, anecdotal evidence was said to indicate a strong link between *merantau* behaviour and the family life course (1998: 50).[21]

While the notion of the life course or life cycle is useful in understanding the restrictions and opportunities that are open to women and men at particular times in their lives, it is also important not to overlook the importance of wider historical processes in changing the canvas on which people's lives are inscribed. As has already been emphasised at several points in this book, Southeast Asia's cultures, societies and economies are undergoing rapid and deep change. The opportunities for non-farm work open to young women in Sungai Gajah and Tiuh Baru were not open to their mothers. The jobs, in large part, did not exist and prevailing moral norms regarding acceptable behaviour made work beyond the household morally reprehensible in any case. Consumption

[18] Although see the discussion above about grandparents looking after their grandchildren while the parents take up extra-local work.

[19] Reardon makes much the same point in his survey of non-farm employment in Africa (1997: 743).

[20] See the discussion of this pattern of mobility in Radcliffe's (1986) study of the Peruvian Andes. She quotes one of her informants as saying: 'When you're single you can just work in the valley. When you have your children, you can't go anywhere' (p. 40).

[21] See also Hetler's study of circular migration from a village one hour's drive south of Solo in Central Java (1989).

Table 6.2   Characteristics of migrant households in Cinta Karya, South Sumatra, Indonesia

	Engaged in *merantau*	Not engaged in *merantau*
Age of household head*	53.0	40.6
Age of spouse**	43.8	33.7
Head-spouse age difference***	9.0	6.9
Persons in household***	3.5	3.8
Number of children***	2.4	2.5
Average age of children***	12.7	8.8
Years of school – head***	2.5	4.1
Years of school – spouse***	1.8	3.6

* significant at .001
** significant at .005
*** not statistically significant
Source: Leinbach and Watkins 1998: 50

pressures, rising educational levels, changing aspirations, technological advances and structural changes in the economies of the region are all providing individuals, as they pass through their life courses, with different and new opportunities and restrictions from their forebears. Patterns of labour force participation, and much else besides, will change not only through individuals' lives but also between generations. It is for this reason that Monk and Katz (1993) emphasise the need not just to identify the life stage a person occupies but also the cohort to which they belong. 'Behaviours we associate with a particular life stage', they write, 'may more truly reflect the conditions through which a group has lived collectively, such as its access to education, than biological age' (1993: 20).

Conclusion

The changes in the household reflect a complex interplay of cultural and economic influences. Globalisation and national industrialisation are offering rural people – but on an unequal basis – the opportunity to take up non-farm work. The changing aspirations of rural people, and particularly the young, are encouraging them to opt out of agriculture (see Chapter 4). Education and contact with the modern world are stimulating a change in the balance of power in the household between the generations and the sexes. This, in turn, is helping to create the conditions in which young women can oppose their parents' wishes.

In her work on Swaziland, Russell (1993) raises – albeit in a different context – many of the conceptual and practical difficulties that plague work on households and household livelihoods in Southeast Asia. She notes that the 'roof and the cooking pot' are not particularly useful in ascertaining identity. And if the distinctive social and domestic structures of Swazi life are ignored then a number of misconstructions are likely to follow concerning the location and dynamics of poverty and wealth, the role and character of rural–urban relations, and the nature of domestic cycles (1993: 762). She also writes about the 'empirical untidiness' that arises out of the interpenetration of rural and urban, and the permeability of household boundaries (1993: 764)

It is difficult to generalise about the trajectory of change in the composition of rural households, and their operation as social and economic units. The evidence from detailed case studies shows considerable variation between areas and countries, even between households in a single village. This is to be expected. Individuals and households juggle available opportunities according to the resources at hand, including land, skills (education) and labour resources. These, in turn, relate to household/family life cycles and to seasonal cycles associated with the demands of agriculture. And behind this layer of explanation lies the political economy of industrialisation and the broader canvas of social and cultural change.

It is out of this complexity that explanatory richness arises. So, while economic and demographic indicators can provide some idea of the direction of change (see the previous chapter), they cannot account for or explain that change, or reveal the diversity that lies behind the broader patterns. This must be sought in the accounts of the lives of individuals and households who represent the mediators through which global and national processes are articulated.

Chapter 7

Agriculture's place in a diversifying rural world

Introduction

The shift from farm to non-farm described in Chapter 5, and the social changes that underpin this shift set out in Chapters 4 and 6, have had profound implications for agriculture. In many quarters these changes have been characterised in negative terms. Land has been abandoned or, at best, a process of disintensification has been set in train. Labour shortages have undermined cooperative work arrangements and allowed productive land to suffer degradation through lack of upkeep. Mechanisation has snatched traditional employment opportunities from marginal groups, forcing them to leave the countryside to survive. Women have seen their jobs evaporate as agriculture is 'masculinised'. And elderly people have found themselves struggling to maintain their farms as young people depart for the city and leave households and villages short of human resources and the vitality of youth.

Agriculture is, it would appear, in the throws of extensive change. In Peninsular Malaysia, Luzon in the Philippines, areas of Java, and across many parts of the Central and Northern regions of Thailand these changes are comparatively far advanced. Elsewhere agriculture remains 'traditional', to the cursory eye at least. A central challenge is to ascertain whether, from the bewildering range of case studies and evidence to hand, any general propositions can be made regarding the nature and direction of change in agriculture. As should become clear as the discussion that follows unfolds, while there are critically important issues of locality which bring a unique character to each and every case study, there is also an underlying current of change which, in terms of direction and constitution, resonates across the region.

A view to the past: change in Ban Ping

Michael Moerman first visited Ban Ping in Northern Thailand in 1959 when, as a PhD candidate he joined a missionary couple and journeyed for two-and-a-half days along a dirt track to get to the village in the district of Chiang Kham, now part of Phayao province (but then part of Chiang Rai province). In his 1968 account of Ban Ping, which is based on this first period of fieldwork, supplemented by a further visit in 1965, Moerman writes:

> In Ban Ping, rice is the main component of every meal, the major source of cash, and the object of most labor. Its production, consumption, and sale are the most common topics of village conversation. All other activities – economic, political, religious, and social – must yield to the rice cycle and the rains that govern it. Rice is a universally accepted standard and store of value (1968: 10).

He returned to Ban Ping in 1979 and again in 1986. At each visit he witnessed a village which, in a spectrum of ways, was undergoing deep and extensive change:

> Ban Ping has withdrawn from farming. Twenty-eight years ago [in 1959], it was a village of rice farmers. Ninety-one percent of village households supported themselves by managing farms. In 1986, only one household in three farmed. . . . Farming, once a way of life, is now a livelihood, one occupation among others (Moerman and Miller 1989: 308).

Behind this shift in the structure of the village economy (and society) in Ban Ping were a host of other agricultural adaptations which individually may not have always amounted to much but, taken together, represent a transformation in agriculture and agricultural production. While it is possible to interpret such adaptations as products of changes in the availability of land, labour and capital, such a mechanistic approach to explanation is inadequate in a number of respects. In particular, it ignores the cultural and social contexts within which economic change is embedded and, furthermore, abstracts economic change from the politics of production. So, while this chapter is concerned with change in farming systems in the region it is also concerned with embedding these changes in the wider political and social milieu and, equally importantly, in assessing to what extent this milieu is, itself, undergoing profound change.

Agricultural labour shortages in an overpopulated region?

> It is ironic for the farmer now – while the farmland is getting smaller, and the population is getting larger, why is there such a shortage [of labour]? Now there's no one to harvest, no one to plant (a farmer in Cavite, Luzon, the Philippines, quoted in Kelly 1999b: 64).

In a region which has seen its population expand from an estimated 32 million in 1800 to more than 500 million today, it may seem perverse to write that labour shortages are highlighted in many accounts as being one of the key factors driving contemporary agricultural change.[1] Such a perspective appears more perverse still when applied to rural areas of high population density like Java, the Central Plains of Thailand, and Luzon in the Philippines. Java's population, for example, has grown by a factor of 23 from an estimated 5 million in 1800 to almost 115 million in 1995. The acceleration in the growth of the island's population dates from the end of the eighteenth century (Figure 7.1). Yet even in areas where agricultural population densities are high and may exceed 500

---

[1] For discussions of Southeast Asia's population and demography see Brown (1997: 81–95), Reid (1988: 11–18) and Owen (1987).

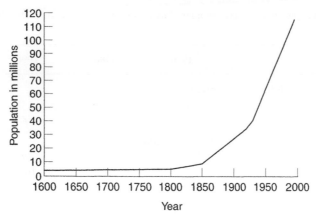

Figure 7.1   Growth in Java's population (1600–1995)

Sources: Based on Reid 1988, Ricklefs 1981, UNDP (1998)

people per square kilometre, local-level studies often paint a picture of severe labour shortages in agriculture. This applies, for example, to work on Thailand (Chantana Banpasirichote 1993, Funahashi 1996, Tomosugi 1995), the Philippines (Kelly 1999a and 1999b, Estudillo and Otsuka 1999, Otsuka *et al.* 1994), Malaysia (Kato 1994, Courtenay 1988), and Indonesia (Naylor 1992, Preston 1989).

Estudillo and Otsuka's (1999) longitudinal study undertaken in Central Luzon, the Philippines (see page 78 for background) offers an insight into the technological changes that have occurred in rice cropping (Table 7.1). Many of these are labour-saving innovations. Furthermore, they argue that the time lag between the adoption of modern varieties of rice (MVs) and the adoption of labour-saving machines 'strongly suggests' that the adoption of MVs was not causal in the adoption of machines (1999: 504). The dissemination of MVs increased labour input in crop care so that actual labour input per hectare remained roughly constant between 1966 and 1994. In effect, the labour-using effects of MVs were cancelled out by the labour-saving effects of machines and direct seeding (1999: 505). What is most striking is the dramatic increase in agricultural wage rates between 1990 and 1994 (Table 7.1). The authors state that this increase cannot be explained except in terms of the sharp increase in labour demand in the non-farm sector (see also Kelly 1999a and 1999b).

Estudillo and Otsuka's perspective holds that changes in the agricultural system are being driven – to a significant extent – by forces that have their origins outside farming and, in many cases, beyond the village. Kelly's work (1999a) in the province of Cavite, south of Manila, also emphasises the decisive links between changing agricultural methods and the wider rural (and national) economy. During the 1990s the number of non-farm jobs in the local area multiplied enormously with the expansion of the Cavite Export Processing Zone (CEPZ). From a handful of small factories employing just 100 people in 1986, the zone hosted 144 factories and a labour force of 38,000 by 1995 (Kelly 1999a: 293), rising to nearly 55,000 in 1998 (Kelly 1999b: 62). Young

Table 7.1   Changes in rice technologies, labour inputs and labour costs, Central Luzon (1966–94)

	1966	1986	1990	1994
Adoption of rice varieties (% adopters)				
TVs	100	0	0	0
MVs	0	100	100	100
Adoption of labour-saving technologies (% adopters)				
Tractors	11	70	95	100
Threshers	77	97	100	100
Direct seeding (broadcasting)	0	9	24	27
Use of herbicides (kg/ha)	0		26	41
Labour input (mandays/hectare)				
Total	64	60	63	71
Hired	40	46	47	49
Real wages*				
Land preparation	100	158	135	277
Transplanting	100	139	131	226
Harvesting	100	151	146	168

Note: * Deflated by nominal paddy price index
TV = traditional variety
MV = modern variety
Source: adapted from Estudillo and Otsuka 1999: 503 and 505

people, for a variety of economic and non-economic reasons were attracted into such work, depriving farming families of labour. The knock-on effects of this shift of young people from farm to non-farm were considerable and wide-ranging.

To begin with, the additional income from such work negated the need for other household members who remained on the farm to take on agricultural work, removing them from the agricultural labour force. Shortages of labour led to a sharp increase in farm wages. In 1995 the wage for a day's planting in lowland Cavite was 100–120 pesos, compared with a national norm of 40–50 pesos. Gangs of planters and harvesters from surrounding provinces were enticed into the province to take advantage of the high wages.[2] But even so it was not unknown for paddy to rot in the fields, such was the shortage of labour. Furthermore, Kelly identifies changes in the sexual division of labour in agriculture, precipitated, he suggests, by the seriousness of the situation (see below). Finally, Kelly interprets the apparent tardiness of farmers in taking up the cultivation of high-value crops like muskmelons (as promoted by the government) in terms of the labour situation. Without available household

[2] Tomosugi in Thailand also writes of a local labour shortage sucking in workers from beyond the immediate vicinity of the village (1995: 65).

labour resources and unable to justify employing wage labour at such high daily rates farmers were continuing to cultivate rice or leaving land idle.[3]

In his study of nearly 100 villages in Northeast Thailand's province of Yasothorn (see page 91), Funahashi (1996) similarly notes the widespread emergence of labour shortages at key points in the rice cycle. The author links this both to the absence of young people and to important changes in the practice of rice cultivation. Formerly, farmers would plant a variety of local and traditional rice varieties with different maturation periods and dates. But, increasingly, just a handful of modern varieties are cultivated, causing labour peaks to become accentuated. The common response to these labour shortages driven by the twin effects of migration and labour demand concentration has been to embrace mechanical land preparation, adopt direct seeding over transplanting, and/or to hire labour to meet any labour shortfall.

The same story emerges in Chantana Banpasirichote's investigation of the village of Klong Ban Pho in Thailand's province of Chachoengsao. In the past rice farming in this village was constrained by a shortage of credit and land. Now the chief constraint, he suggests, is a lack of labour (1993: 13). Like other studies noted here, changes in agriculture are seen to be a response to emerging and intensifying shortages of labour, forcing households to make adjustments to their farming systems. Buffaloes have been replaced by ploughing machines, and when the cost of labour reached 100 baht per day (at that time, equivalent to US$4), households switched from transplanting to direct seeding.[4] Rice harvesting machines and mechanical threshers were also introduced to make up for the increasingly severe labour shortfalls (Chantana Banpasirichote 1993: 14).

Finally, as illustrative of the range of responses that farmers have made in the face of labour shortages is Preston's study of Central Java, significantly titled 'Too busy to farm' (1989). He describes how villagers have replaced vegetables with fruit trees in their home gardens, shifted from the annual cultivation of some crops to their cultivation in alternate years and, in some cases, curtailed the traditional practice of dry season cultivation of paddy land with non-rice crops (Preston, 1989).

These accounts – and there are many more which tell a broadly similar story – link labour shortages to the competition for labour between farm and non-farm. But on paper, and even with the proliferation of non-farm opportunities detailed in earlier chapters, there should still be ample labour to meet the demands of agriculture in most areas of the region. Factories may have selectively removed some people from the agricultural labour market, but not on a sufficient level to create the widespread shortages noted in the literature. To square the circle it is necessary to look to the accompanying social and cultural changes. Taking this perspective, the short answer is not that there is an

---

[3] This sequence of change, with regional variations, is corroborated in the work of Douglass on Indonesia (1998: 19–20).

[4] Miyagawa (1996: 36) notes the same for Ban Don Daeng in Khon Kaen province, where the cost of farm labour increased from 20–25 baht in 1983 to 50–70 baht in 1991, stimulating, in his view, the switch from transplanting to direct seeding. See also Konchan and Kono (1996) on the spread of direct seeding in the same region of Thailand.

Box 7.1 From transplanting to direct seeding

Of all the changes in rice cultivation, perhaps the one most clearly indicative of the changing economic conditions in the countryside is the widespread shift from transplanting to direct seeding (broadcasting).[1] As the former is arguably the world's most productive – in terms of land – agricultural system, the shift from transplant rice culture to direct seeding appears, to many, a retrograde step and one which goes against the evolutionary 'trend'. However, a combination of economic, cultural and technological changes have both propelled and permitted this change. To begin with, and as noted in this chapter, quite severe labour shortages have emerged in many areas as rural people have been enticed into non-farm work. Second, and also noted here, changing cultural preferences have created a climate where farming is avoided by many young people. Together, these changes have encouraged farmers to save labour wherever possible. A range of studies show that direct seeding requires about half the labour necessary for transplanting (Pandey and Velasco 1999:8). But advances in technology, and particularly the development of chemical weed control methods and high-yielding, quick-maturing rice varieties have provided the technical means to embrace direct seeding. The shift from transplanting to direct seeding varies across the region and is most pronounced in Malaysia, the Philippines, Thailand and Vietnam.

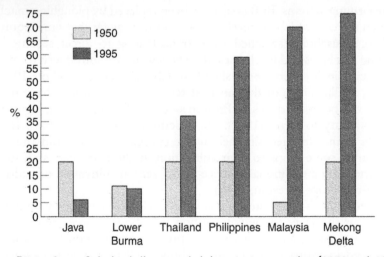

Figure B7.1 Area of riceland direct seeded, by country or region (1950 and 1995)

Source: Based on Pandey and Velasco 1999

[1] There are two forms of direct seeding: dry seeding and wet seeding. Dry seeding involves broadcasting (or dibbling or drilling) the ungerminated seed onto prepared dry soils. Wet seeding involves the sowing of pre-germinated seeds onto prepared, saturated soils.

---

absolute lack of labour in rural areas to meet farming needs, but rather a lack of labour *willing* to work in agriculture. Young people increasingly view farming as a low-status occupation to be avoided (see page 53) and women from wealthier households have, in some instances, retreated into the house. Kelly's (1999b) survey of two villages in Cavite noted above revealed that almost one-fifth (18 per cent) of those aged 15 to 60 years old had no direct source of livelihood and were 'standing by', effectively waiting for an acceptable job opportunity to arise. Furthermore, of this fifth more than a half were aged 15–30 years old. In other words, in Cavite severe shortages of labour in agriculture are accompanied by apparent underutilisation of labour. This point is also made by Hart in her study of the Muda area of Malaysia (Hart 1992).

How labour relations in agriculture are evolving in Southeast Asia is clearly linked to much more than just the availability and cost of labour. As the discussion in Chapters 4 and 6 made clear, the reconfiguration of the household and its operation, changing aspirations – especially among the young – and altering relations between the genders and generations are also critical in understanding the changes that are taking place in farming systems in the region. Adding a cultural piece to the jigsaw does help embed an understanding of agricultural change within the wider context of social change and the shifting priorities and aspirations of people in the region. Yet even this is inadequate in that it tends to ignore the political economy of agricultural change.

Mechanisation: competing interpretations of the Muda experience

The debate over the mechanisation of agricultural production exemplifies the complex explanations that have been employed to account for different facets of agricultural change. And, arguably, nowhere in Southeast Asia has the debate

Plate 11   Threshing rice by hand in Mahasarakham, Northeast Thailand (1982).

Plate 12   Threshing rice using intermediate technology, the highlands of Toraja, South Sulawesi, Indonesia (1991).

Plate 13   Mechanised threshing, Phrae, Northern Thailand (1992).

Plate 14   Combine harvesting, Chiang Mai, Northern Thailand (2000).

over mechanisation been richer than for the Muda area of Kedah in Peninsular Malaysia (see: Scott 1985, Hart 1992, Wong 1987, De Koninck 1992, Barnum and Squire 1979).

The state-directed development of the Muda plain for agriculture can be dated from the seventeenth century when, under the orders of the rulers of Kedah, several large canals were constructed using corvée labour (Overton 1994: 53, Bray 1986: 98–100). Indeed, the sultanate was an important rice exporter before the advent of the colonial period in the Peninsula. Additional canals

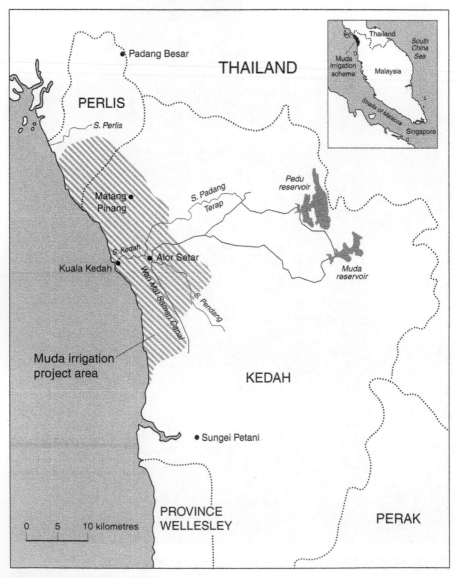

Figure 7.2 The Muda Irrigation Scheme, Kedah, Peninsular Malaysia

Source: Based on De Koninck 1992, MADA 1981

were built in the early years of the twentieth century, following the imposition of indirect British rule in 1909. However, the decisive shift from labour-intensive, transplant rice agriculture to mechanised production can be linked to the initiation of the Muda Irrigation Project. This involved, with World Bank support, the construction of two large dams along with associated headworks, main canals and secondary canals, and an institutional infrastructure provided by the Muda Agricultural Development Authority (MADA) (Scott 1985: 64–5). The initial plans were drawn up between 1961 and 1964, the main period of construction spanned the years 1966–70, and the scheme became fully operational in 1973, by which time double cropping of rice was possible on 96,000 hectares of land (Figure 7.2). In 1974 the Malaysian government and the World Bank hailed the scheme as an outstanding success: 92 per cent of the project area was being double cropped, MVs almost universally planted, and the region was producing a large rice surplus to contribute towards Malaysia's drive for national rice self-sufficiency (Scott 1985: 65). Production increases were 'spectacular', rising from 268,000 tons in 1965 to 678,000 tons in 1974 (De Koninck 1992: 28). Less expected was the rapid mechanisation of production that followed close on the heels of the spread of double cropping. By the early 1980s almost the entire Muda area was being harvested using combines.

The neo-classical explanation for the spread of labour-saving innovations like the combine assumes that mechanisation was a simple price-induced response to a shortage of labour. In some accounts (e.g. Barnum and Squire 1979) labour supply, constrained by a growing preference for leisure, failed to meet the increases in labour demand generated by the green revolution. In others, labour shortages are more particularly linked to the increase in non-farm opportunities for young women and men. However, in both instances, the introduction of combine harvesters is seen as a classic case of induced innovation driven by simple relative scarcities of factors of production. Figure 7.3 shows the pattern of labour use in wet rice cultivation and the adoption of

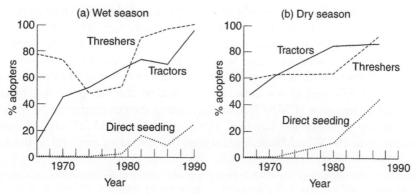

Figure 7.3 Labour use and adoption of labour-saving innovations in wet rice agriculture, Central Luzon (1966–90)

Source: Data from Otsuka *et al.* 1994

Table 7.2   Distribution of labour arrangements, Muda, Kedah, Malaysia (1970–79, %)

	1970	1974	1979 Transplanting	1979 Harvesting
Family	34	39	26	16
Hired	32	59	69	70
Exchange	34	2	5	14
Total	100	100	100	100

Source: Hart 1992: 814

labour-saving innovations based on a longitudinal study from Central Luzon. Labour use per hectare increased with the introduction of the green revolution from the mid-1960s, and then declined as the adoption of labour-saving innovations including machines and direct seeding accelerated from the late 1970s.

A second line of explanation, most clearly linked with the work of Gillian Hart (1992), is what she terms a 'politicized approach'. This 'helps to clarify a key dynamic at work in the Muda countryside during the 1970s – namely, an intensifying set of struggles over labour mobilization and control . . .' (1992: 811–12). Women, and especially poorer women, collectively organised to push up wages, thereby pitting themselves against large (male) landowners. The explanation for the rapid mechanisation of production, in Hart's view, therefore lies in the conflict between poor(er) female workers and rich(er) male landowners.[5] During the 1970s traditional forms of reciprocal labour exchange in the Muda area were replaced by wage labouring systems in which gangs of workers would move from farm to farm to transplant and harvest rice. In 1970 for the Muda area as a whole approximately one-third of labour was hired. By 1979 this had increased to more than two-thirds (Table 7.2).[6] Women resisted attempts by landowners to break up the gangs, refusing to work on an individual basis even when demand for labour was depressed. De Koninck describes how some gangs would refuse to transplant for those farmers who would not also agree to employ them for harvesting (1992: 110). This conflict between landowners and labour was also, to a significant degree, a conflict between men and women. Male labourers did not organise, a fact that Hart puts down to patronage relations in the area and the role of the ruling United Malays National Organisation (UMNO) in implementing the government's New Economic Policy (also see De Koninck 1992). Moreover, MADA officials, with the support of the World Bank, encouraged the spread of mechanical innova-

[5] She also notes the failure of neo-classical interpretations to take account of the operation of the household economy and the different sets of rules governing the behaviour of unmarried sons and daughters. This issue was discussed at some length in Chapter 4.
[6] Tomosugi also noted the disappearance of such traditional forms of labour exchange: 'Rice cultivation is now managed like a modern business. . . . The relationship between farmers and labourers is mainly stipulated in terms of money, not by traditional intimacy' (1995: 65).

tions, assuming labour shortages were acute. Hart contends that MADA were intent on removing farmers with sub-viable land holdings from the area, believing that mechanisation (along with direct seeding) would assist in the process.[7]

The ultimate failure of female work gangs to prevent the mechanisaton of production in Muda, perhaps always futile, can be linked to a number of factors. The spread of direct seeding undermined the bargaining power of transplanters; conflicting intra-household interests pitted women (wives and daughters) against men (husbands); competition between gangs from different villages undermined inter-community solidarity; and the wider political economy created conditions favourable to mechanisation by reducing the relative price of capital to labour.

Depending which interpretation is embraced, the social effects of mechanisation take on a very different hue.[8] It is possible to argue, using the political economy approach, that mechanisation displaced poorer women from productive work, forcing them into an increasingly marginal position. Such a view would also resonate with the 'masculinisation' of farming perspective, in which women are excluded, leaving agriculture a largely male preserve (see the next section). Men controlled the use of machines and also the few manual jobs that remained after mechanisation – such as the bagging and transport of rice from the fields to the roadside (De Koninck 1992). Furthermore, the substantial growth in non-farm activities during the 1980s was largely due to the increase in women's participation and this, too, 'can be interpreted as a consequence of the marginalisation of their skills in the padi production process' (De Koninck 1992: 177). The neo-classical interpretation, on the other hand, would see mechanisation as contributing to and supporting agricultural production in a context of labour shortages. Moreover, implicit in some neo-classical studies is the contention that the introduction of farm technology will have a redistributive or, at worse, a neutral effect on non-adopters. Using data from the Muda area, Barnum and Squire, for example, argue that agricultural projects 'which benefit directly those members of rural society who control the main productive assets, can also be expected to benefit those who own little or no land through changes in wage-rates and the wage bill' (1979: 99). This, of course, is fundamentally at odds with the interpretation offered by Hart (1992) and Scott (1985).

Gender divisions of labour in agriculture

Studies on changing gender divisions in agriculture have variously argued that agriculture is being 'masculinised' (e.g. Parnwell and Arghiros 1996: 21–2) and 'feminised' (e.g. Trankell 1993: 82). The important point, however, is that the changes outlined here differentially impact on men and women and, in some instances, are helping to remould hitherto established and entrenched gender divisions of labour.

---

[7] The important issue of the persistence of sub-livelihood landholdings in the face of technology-driven agricultural change is discussed in Chapter 8.

[8] This mirrors, but on a smaller scale, the rather broader question of the trajectory of agricultural development in the region and whether it is, in essence, developmental or not (see page 88).

In Sagada in the Cordillera of Luzon there have been a sequence of changes in agriculture which have variously impacted on the gender division of labour (Preston 1998: 372–7, Jefremovas 1992). Traditionally, swidden farming of sweet potatoes on the hillsides was complemented by rice cultivation in the valleys. In both systems, women would provide most of the day-to-day labour – although men cleared the land for swiddening – and they were also the guardians of specialised agricultural knowledge. Swidden farming of the hillsides began to decline after 1945 and has now been virtually abandoned. Sweet potatoes, the traditional swidden crop, are no longer in demand and the loss of men to migration has reduced the size of the available labour force. Sweet potatoes have been replaced by pine forests, an attempt by farmers to secure their ownership of the land while disengaging from annual cultivation of the hillsides.[9] Instead households have concentrated their time and effort on the cultivation of vegetables for sale. In some cases vegetable cultivation of the valleys has displaced (subsistence) rice agriculture altogether. This sequence of changes has had important effects on the gender division of labour in agriculture. The freeing of men and women from swidden agriculture has permitted men, particularly, to concentrate their efforts on vegetable cultivation. While women do contribute their labour during peak periods, vegetable farming has become a largely male preserve. Women continue to cultivate the rice lands but these have decreased in extent as vegetable cultivation has expanded onto traditional rice lands. Preston also links these changes with the adoption of values prevailing in urban Philippine society (1998: 377): the preference for rice over sweet potatoes, for urban work over agricultural work, and for cash-generating vegetable cultivation over the exhausting business of cultivating the hillsides. But while the division of time between men and women in different agricultural activities may have changed, Jefremovas contends that this should not be interpreted as a loss of power or influence. Financial decisions are made collectively, women continue to hold the household money and dominate trading and marketing, and household decisions regarding cropping strategies and sales are also arrived at jointly (1992: 53–4).

That agricultural change has different impacts on women (and men) in different areas is exemplified in the contrasting interpretations offered by accounts of the Muda area and Luzon. As described above, labour shortages and mechanisation in the Muda irrigation scheme excluded women from productive work. As Scott says of the village of Sedaka, the combine became popularly known as *mesin makan kerja* – the 'machine that eats work' (1985: 154) – and its introduction from mid-1976 was an 'economic disaster' for poorer households composed of pure wage labourers (p. 116). The effect of the combine was to sever the economic connections that hitherto had linked rich and poor in a mutually dependent (but unequal) relationship. Women from wealthier households retreated into the household as part of a process of housewifisation or domestication propelled, at least in part, by government policies which see farming as 'men's work' and by a culture of modernity which places women

---

[9] Swidden land traditionally belonged to the farmer for only so long as it was farmed.

firmly in the domestic sphere (Ariffin 1992, see also King 1999). Blackwood, with reference to a Minangkabau village in West Sumatra, writes of the 'discourse of the domestic' whereby government policies have tried (not always success-fully) to embed women in the household, as wives, mothers and home-makers (1995: 135–8). But while women from wealthier households were transformed into 'princesses of the house', poorer women found themselves marginalised and forced to look for employment beyond the village – a 'haemorrhaging' of the poor, as Scott puts it (1985: 124). In his work in Luzon, Kelly also identi-fies changes in the sexual division of labour in agriculture precipitated by the seriousness of the labour shortages brought about by the withdrawal of people (for a variety of reasons) from agriculture. But in this instance, women were not displaced from agricultural work as a result of mechanisation. Instead they – indeed whole families – were drawn into field work, whereas previously women would be confined to marketing and financial management (Kelly 1999a: 294). At a general level it is possible to state that members of poorer households, pressured by the exigencies of survival, regularly and persistently cross culturally set gender boundaries while those of richer households are able to observe such boundaries (see Elmhirst 1997 on Sumatra, Indonesia and Zaman 1995 on Bangladesh).

Agricultural decline and the generations

An important question concerns whether the changes outlined here and in earlier chapters represent the abandonment of agriculture, or the social reconfiguration of production. Does the predominance of older cohorts in agriculture – the geriatrification of farming – mean that when these people retire or die agriculture will cease; or is it simply that people are delaying their entry into farming until middle age?

Most non-farm opportunities are selectively embraced by younger men and women. This is both because younger people are in a stronger position to exploit such opportunities (they are usually unmarried and childless and more mobile and are also more likely to have the necessary educational qualifica-tions) and because employers prefer to hire younger (unmarried) men and women, and especially women. But even with the attractive wages that non-farm work may bring, along with the status that can accrue from working in the modern economy, it is nonetheless true that such work is often unreliable and sometimes marginal (as the 1997–99 economic crisis exemplified at a regional level). Mills writes in her work on factory workers from rural back-grounds in Bangkok: 'rural women realize that a return to the village house-hold remains their most reliable source of economic security after marriage' (1997: 48). But she also notes that many 'hoped to be more than just "rice and upland crop farmers" (*chaaw naa chaaw ray*) when they left the city' (1997: 49).

Some scholars suggests that this is not a life cycle shift but a permanent change in the complexion of the rural economy (e.g. Ritchie 1993: 13, Chantana Banpasirichote 1993: 38, Hart 1992). What, they imply, will happen when middle-aged farmers retire? Who will fill the void? Putting aside, for the moment, the issue of whether those people who have left the countryside will return (and if they do, whether they will return to farming), there is also the question of whether young people have the knowledge and skills to farm. In the 'worst cases' farming knowledge has not been passed on to the younger generation and they do not know how to farm (or to manage a farm, which is rather different). The arts of transplanting and harvesting, of water management, of seed selection and of land preparation, are all lost to this generation (Ritchie 1993: 14, Chantana Banpasirichote 1993: 11–12). But, significantly, it is rare to find examples of parents lamenting this loss of knowledge. For them, as for their children, the future of work is seen to lie elsewhere.

Hart is another scholar who considers the future for agriculture, in this instance in the Muda area, to be dim. She finds it hard to believe that when the current generation of farmers die their sons, employed in well-paid non-farm jobs 'could be persuaded to transform themselves into the yeoman farmers envisaged by the World Bank' (1992: 820). In all likelihood, she hazards, farming will continue on a part-time basis and cultivation practices will continue to decline.

Studies of farm life in Southeast Asia, and in most other parts of the world, have made much of the attachment that farmers have to their land. This is often described in almost spiritual terms, as if the loss of land would be akin to the loss of soul, and certainly of identity. However, it is worth considering whether such attachment is immutable, indeed whether it was ever quite as elemental as implied. The processes and changes discussed in earlier chapters hint at a disintegration of the bond between people and the land, and especially among the younger generation.

'Squeezing' agriculture

Hub Takhood is a coastal village close to Khao Sam Roi Yod National Park in the province of Prachuab Khiri Khan, about 275 kilometres south of Bangkok. In 1972 a road was constructed to the village as part of Thailand's Accelerated Rural Development programme. For the first time Hub Takhood was well connected with the outside world and by the end of the 1970s prawn farming had become the economic mainstay of the village economy, taking over from subsistence rice farming and coconut production (Hirsch 1992: 45–7). But it was the expansion of prawn farming itself which caused rice farming to 'all but cease' in the area. The removal of protective mangroves to make way for further prawn ponds increased salinity in fields that were already saline, making them unsuitable for cultivation using even the most salt-tolerant varieties of rice.[10]

An economistic interpretation of agricultural developments in Hub Takhood would stress the extent to which farmers moved from rice cultivation into an activity – prawn farming – offering far greater returns. But the economic

[10] Also see Flaherty et al.'s (1999) paper on inland shrimp farming in Thailand and its environmental effects.

changes in Hub Takhood need to be understood in terms of the space that capital and the state provided. For they were also instrumental in shaping the pattern of development in the area. The state restricted (for environmental reasons) villagers' use of the uplands while (mostly) ethnic Chinese middlemen and commercial banks provided the critical financial ingredient. Prawn farming, and especially semi-intensive prawn farming, requires considerable initial investment. In the mid- to late-1980s this amounted to 100,000 baht (US$4,000) per *rai* (6.25 rai = 1 hectare), or US$80,000 for a farm of 20 *rai*. In 1986/87 the average farm income of households in the South was just 20,500 baht (MOAC 1989). No local family could consider prawn farming on this scale without the involvement of outside investors. 'Capital', Hirsch writes, 'effects control either over land itself or the way in which it is used' (1992: 48).

Furthermore, the pattern of development in Hub Takhood would seem to have marginalised poorer households. Rice farming has been undermined by increased salinisation; crabs and other marine life have disappeared along with the mangroves; uplands, formerly used for the collection of non-timber forest products are off-limits as the state has intensified its control over forest land; and cropping of upland areas has been barred for similar reasons. It would seem, in the case of Hub Takhood, that not only have traditional agricultural pursuits been squeezed pincer-fashion by capital and the state, but poorer households have been squeezed out of agriculture and the local area. Hirsch (1992: 47) reports that some villagers work in a pineapple canning factory in nearby Pranburi but otherwise there are few locally available opportunities for villagers.

The environmental, and thus the agricultural, impacts of the interpenetration of urban and industrial activities into rural areas have been noted from several areas. In Cavite, south of Manila in the Philippines 'farmers complain that irrigation canals have become silted up with eroded material from local building sites . . . [and] . . . that crop pests have become an increasing problem with the development of residential areas in the midst of farmland' (Kelly 1998: 43). Field bunds and dikes have been broached, drainage systems disturbed by the construction of roads and expanses of concrete, rats (a serious agricultural pest) have multiplied as residential areas have grown, waste and refuse have found their way into water courses and fields, and the delicate balance that maintains the rice field ecosystem has been disturbed, undermined and, in some cases, irreparably damaged (Kelly 1999a: 296–7).[11]

Land use conversion creates a cycle of environmental change and agricultural decline. Farmers abandon their land due to declining productivity brought about by urban/industrial-induced environmental change; their land either stands idle, increasing pest infestation or is, itself, sold to property developers or industrial interests, further undermining the environmental integrity of the

---

[11] There are many similar examples of environmental decline initiated by the infiltration of non-farm activities into rural areas. Kirkby and Zhao note extensive pollution of farmland by Township and Village Enterprises (TVEs) in China (1999). Reichert writes how the demands of the expanding brick-making industry in the Nile Valley of Egypt has removed the top soil from some of the best agricultural land, adversely affecting production (1993: 50).

land. These environmental changes and their concomitant agricultural impacts need to be seen against the forces described in the previous section. For example in the study of Klong Ban Pho noted above, the author writes that 'rural industrialization, as indicated by the penetration of factories into rural areas, has challenged the agricultural sector, which is experiencing declines in productivity' (Chantana Banpasirichote 1993: 33). This has accelerated the process of occupational shift in the countryside which, in turn, has encouraged the labour shortage-induced agricultural changes. But elsewhere, as in Hub Takhood, the decline of agriculture is better explained as part of a destructive process engendered by the environmental conflicts between different activities. While the effects of land degradation may be widespread, it has been suggested that the 'vortex' of poverty and need which forces poor households, especially, to embrace non-farm activities can lead to particularly acute conflicts of interest. Bryceson, with regard to Africa, notes that non-agricultural activities may be pursued by the group to the detriment of the agrarian base (1997b: 239).

Complementary or conflicting? Emerging rural–urban relationships

The discussion here affords the impression, as noted in the introduction to the chapter, that the effects of intensifying rural–urban relations are deleterious to agriculture. But mechanisation and the introduction into rice cultivation of various labour-saving technologies and practices, from rotavators to pre-emergence herbicides and direct seeding, do not, in themselves, necessarily herald or indicate a decline of agriculture. Nonetheless, many studies do make this assumption, seeing farming entering a period of crisis fermented by a combination of low returns, competition from non-farm activities, environmental degradation, and lack of interest among the younger generations. However, a decline in the productivity of agriculture cannot be equated with a decline in livelihoods. It should be remembered that it has been the failure of farming to meet the social and economic expectations of rural people which has fuelled these changes. To be sure, the political economy of rural development has played a role in undermining agriculture, but this should not detract from the reality that for many households farming offers few prospects for a better life. Furthermore, agriculture must be viewed as being involved in a dynamic relationship with non-agriculture. For Kelly (1998, 1999a) the outcome of this dynamic relationship is the 'squeezing' of agriculture; for others there is a creative interlocking of farm and non-farm livelihoods(see Kamete 1998). The question of whether there is a 'virtuous cycle' of farm/non-farm relations or a destructive one is discussed in more detail in Chapter 8.

Land abandonment and the end of agriculture?

It would seem that the most striking evidence of the changing economic priorities and opportunities in some areas of rural Southeast Asia lies in the abandonment of agriculture and the increasing prevalence of idle or abandoned land. However, the presence of such land may be the product of a number of factors and not necessarily indicative that the returns to agriculture are so low that farmers have simply shifted their labour into other, non-farm activities.

To being with, idle land may be a function of speculation rather than a lack of interest in farming. The prevalence of land speculation in some areas of Southeast Asia needs to be seen in the context of the role of the state in allowing individuals to secure ownership over land. In Thailand, the state encouraged farmers to claim title to their land. As roads made formerly remote areas accessible, so land became an increasingly valuable resource to non-local people intent on its conversion to other uses. Young villagers found they could no longer buy land to farm because of its rapidly escalating price and were forced out of the village to find work (whether willingly or unwillingly). 'The "simple" act of building roads', Ritchie writes of the North of Thailand, 'has complex repercussions sociologically and economically . . . by providing easy access to land which was formerly remote . . . [enabling] land speculators to go out into rural areas' (1993: 6). The price of land in Ban Lek, the village in Northern Thailand where Ritchie worked, increased by 2,300 per cent between 1985 and 1991 (1996b). In the Philippines, the legal framework of the Comprehensive Agrarian Reform Law (CARL) of 1988 also encourages landowners to hold land idle for reasons of speculation. Under CARL, allowing a tenant to cultivate the land makes it less likely that a non-agricultural zoning status will be approved by the local authorities. Furthermore, the longer a tenant cultivates a piece of land the higher the compensation due to him or her. Thus, families will sometimes hold onto land but leave it idle rather than rent it because, first, of the risk of alienation by a tenant farmer and, second, because it permits the owner to apply for non-agricultural land zoning status (Preston 1998, Kelly 1998: 45 and 1999a: 293–4).

Box 7.2   Farewell to farms

Many of the points made in this chapter resonate with the general tenor of Bryceson's (1996, 1997a) and Bryceson and Jamal's (1997) work on Africa. The title of the latter, edited volume is 'Farewell to farms: de-agrarianisation and employment in Africa'. To begin with, Bryceson challenges the assumption that de-agrarianisation – which she argues is an Africa-wide phenomenon – is a negative process. The easing of people away from a strictly agararian existence is, in their view, 'rational in economic, social, political and psychological terms' (Bryceson 1997a: 3). De-agrarianisation is defined as consisting of four parallel processes: occupational adjustment, livelihood reorientation, social re-identification, and spatial relocation (p. 4). Indeed, such are the attractions of diversification that they they wonder why 'people continue to engage in agricultural activities at all' (p. 9). As in the discussion here, they note the predominance of young people abandoning farming and the strong social pressures on younger generations to avoid agriculture and (by implication) to embrace non-agricultural pursuits. Older generations, by contrast, show a greater commitment to the land, to their natal villages, and to agricultural work.

Figure 7.4 The sequence of agricultural change under conditions of labour scarcity in rice cultivation

Note: this sequence, of course, can be reversed and the trend indicated here is only reflective of the general situation in the region.

When the legislative and political systems favour industrial development over agriculture, land speculation is likely to be instrumental in creating new agricultural geographies. Patterns of land use are shaped, in such instances, less by labour shortages and competition between agriculture and non-agriculture at the household level, than by wider political and legal processes and frameworks. To be sure, there is a tension weaving its way through the political economy of rural–urban relations, but it is a tension which has its origins outside agriculture and beyond the village.

While land speculation may be the most important reason explaining why land is left idle (aside from labour shortages), there are also other factors to consider. Year-to-year variations in climatic conditions can also have a considerable effect by taking some land – particularly if it is marginal rice land – out of production (see Rigg 1985, Grandstaff 1988). Finally, the environmental consequences of the infiltration of industrial and other activities into rural areas can also render land unsuitable for agriculture (see above). But notwithstanding these factors, there is also ample evidence that some land has been abandoned simply because there is not the labour, nor the economic incentive, nor the interest, to sustain its cultivation. At a simplistic level it is possible to set out a sequence of changes of which the abandonment of farming is the last (Figure 7.4).

The country which has seen the most dramatic increase in idle/abandoned land is Malaysia, and particularly Peninsular Malaysia. Kato (1994) explains the prevalence of abandoned *sawah* (wet rice land) in Malaysia in terms of the twin processes of 'de-agriculturalisation' and 'de-kampong-isation'. De-agriculturalisation refers to the lack of interest in agriculture (particularly among the young), coupled with the absence of any need to cultivate land among the older generations due to the abundance of non-farm work and the income that it generates.[12] De-kampong-isation refers to the growing spatial fragmentation of households as people take up work far from the village, and for increasingly long stretches of time. Together, they account for the significant expansion in *sawah* 'covered with bushes and shrubs' (Kato 1994: 168).

Conclusion

Recent work on the local configuration of the labour market has stressed the degree to which labour market segmentation is a response to factors embedded in local social and economic space. To understand how rural people are responding, in employment terms, to the changes and opportunities that confront them is a product of global processes articulated through and within the local social setting. Thus the ability of young women to engage in non-farm work, of men to leave home for extended periods, of sexual divisions of labour in agriculture (should they exist) to metamorphose, and for tenant farmers to take control of land use decisions, are all a reflection of local conditions. 'The demand and supply of labour occurs through a complex local prism of cultural expectations and identities' writes Kelly (1999b: 57) with reference to the Philippines, a view supported in Pincus' study of West Java, where wage labour relations are 'bound up with the unique pattern of village formation in each location' (1996: 136). Moreover, it is not just the political economy of locality which is important in helping to determine patterns of change in any particular area; ecological conditions are also important. Linkages between rural and urban areas, trajectories of change in agriculture and the nature of technological change are all contingent, to varying degrees, on environmental conditions.

A second issue, along with the importance of locality, which comes through in the case studies discussed here is the necessity of looking beyond the economic. Important cultural and social changes are influencing the decisions that rural people are taking, and this applies particularly to the young. Furthermore, the politics of land use, and the manner in which politics intersects with economics, is highly important in, for example, understanding land speculation and the development of new activities in rural areas. It is also the intersection of cultural and social with economic which helps in understanding patterns of industrialisation in Southeast Asia, and the role of rural areas and rural people in that process.

---

[12] Kato (1994) refers to a 1987 village study by Zulkepli Awaluddin in which 61 per cent of respondents said their livelihoods had improved and 32 per cent said they were more or less the same against an agricultural backdrop in which 84 per cent of irrigable sawah stood idle/abandoned.

Chapter 8

Rural industry and farmers in the city

Introduction

In Chapter 2, it was suggested that one of the key reasons why some scholars have demanded a rethinking of the agrarian question is the growing mismatch between spatial categories (rural–urban) and sectoral categories (agriculture–industry). Not only is there a vital human landscape of rural–urban relations, as people oscillate between country and town and between farm and factory, but increasingly the factories themselves are infiltrating rural areas. Moreover, rural entrepreneurs are diversifying into (predominantly) craft-based industries and urban-based enterprises are employing villagers to carry out piecework in their homes. In some cases, craft-based industries are expanding and linking in with global markets, in the process making the transition to becoming small-scale capitalist industrial enterprises. Finally, all this is occurring in a context where urban centres – so-styled extended metropolitan regions (EMRs) – are intruding into and merging with rural areas. In short, people's lives, the articulation of agriculture and industry, and the nature of the rural and urban economies are becoming increasingly complex and interlinked. For Watts and Goodman, studies of rural industrialisation in the developing world and on rural non-farm work represent 'some of the most exciting recent work in agrarian studies' (1997: 16).

There are five avenues by which farmers (or rural people) are being drawn into non-farm (or industrial) work. To begin with, large numbers are moving to urban areas to take up work in the informal sector, often living in burgeoning shanty towns and squatter settlements. Second, there are also significant numbers of people who are moving to urban or peri-urban areas to work in factories or on construction sites, many of which provide accommodation in dormitories or in on-site barracks. Third, factories are being established in rural areas, drawing their labour force from the surrounding countryside (or, in some reported instances, from nearby urban settlements). Fourth, some rural people are establishing small-scale industrial activities *in situ* becoming, in the process, entrepreneurs or proto-industrialists. And fifth, some people in rural areas are engaging in piecework for factories located elsewhere, often in the core urban region.

Plate 15 Traditional forging of knives, Ban Rong, Northern Thailand. Farmers' knives have been made in this way in Ban Rong for more than a century. Technological change has been minimal.

The industrialisation of the countryside

The assumption implicit in much of the literature is that industrialisation, and the wealth and economic vitality that it generates, is centred either within or on the periphery of urban areas. In those instances where rural industrialisation is occurring, it is usually characterised as proto-industrialisation, petty commodity production or, possibly, petty capitalist production – a stage in a historical process towards the emergence of an industrial economy. Rarely are rural areas considered important in the process of industrialisation except in two respects: as pools of cheap labour, and as a source of agricultural produce to support the industrialisation project. However this perspective, which can be linked back to Kautsky's work on the agrarian transition (see Chapter 2), is increasingly out of touch with emerging spatial patterns of industrialisation in the Asian region. As will become clear, in some countries rural areas have become central components in the industrialisation process.

The emergence of so-styled extended metropolitan regions (EMRs, see page 59) has blurred the boundaries between rural and urban areas and, therefore, the distinction between rural industrialisation and urban industrialisation. The greatest academic (and policy) attention has been directed at the key metropolitan centres like Jakarta, Bangkok, Metro-Manila and Ho Chi Minh City. However, the same process, albeit at a lower level, is evident in urban centres with smaller populations and lower profiles. This is true of Medan in Sumatra, Surabaya and Bandung in Java, Chiang Mai, Khon Kaen and Hat Yai in Thailand, and Johor Bahru in Malaysia, for example. Guinness and Husin in their study of the sub-district of Pandaan in East Java write of 'industrial villages' ringing the town of Pandaan, where factories draw their workers from local rural settlements (Guinness and Husin 1993: 282–4). The settlements even closer to Pandaan town have become, in their term, 'urban villages'. These have, over time, been absorbed into the urban fabric and are closely integrated with the town in terms of employment and economy (Guinness and Husin 1993: 285).[1]

This division of the space economy into different categories according to the proximity and level of interaction between rural and urban areas is tempting and experientially persuasive. (Driving out from any one of the towns noted above emphasises the apparent validity of such a pattern of development.) However, the geography of rural–urban interactions, and the intersection of space and activity, not to mention space and humanity, is rather more complex and raises a series of further issues.

Firstly, improving transportation has allowed stages in the production process to be divided between city and countryside (see below). Secondly, the experience of China, especially (see Box 8.1), has emphasised the potential for rural areas to become national engines of economic (industrial) growth *in their own right*.

Box 8.1   Industry in the countryside: rural industrialisation in China

The most telling and impressive example of the role that rural enterprises can play in economic growth comes from China. The figures speak for themselves: 23 million rural enterprises in 1996 employing 135 million workers (roughly one-third of rural China's working population) and contributing 23 per cent of GDP, 44 per cent of gross industrial output value, and 35 per cent of export earnings. As Smyth says, rural enterprises or TVEs (Township and Village Enterprises) have been the 'backbone' of China's economic record in recent years (Smyth 1998: 784) and are a significant element in every province's economy. In some cases rural enterprises are the dominant presence. This is particularly true in coastal provinces, in regions close to urban centres, and in areas where a buoyant agricultural sector supports local markets for non-farm products (Parish *et al.* 1995: 699). In the coastal province of Zhejiang south of Shanghai, for example, 75 per cent of industrial output is generated by rural enterprises (Smyth 1998: 788). In their study of ten counties scattered across the eastern two-thirds of China, Parish *et al.* found that somewhere between one-fifth and two-fifths of people were engaged, primarily, in non-farm work (1995: 705–6). Thus TVEs are not only important in the context of rural China but 'will be a vital factor in the nation's overall development trajectory' (Kirkby and Zhao Xiaobin 1999: 273).

   Compared with the countries of Southeast Asia, the Chinese government has been remarkably successful in limiting the flows of rural people to urban areas – ▶

---

[1] They also distinguish between 'central agricultural villages' and 'remote agricultural villages'.

Box 8.1 (continued)

despite the presence of considerable surplus labour in the countryside. In 1988, for instance, there were estimated to be 100 million surplus farm workers in China, nearly one-third of the total rural labour force. By 1995, Wang and Hu report, this had risen to 150 million (1999: 78).[1] The fear of a massive shift of the rural un- and underemployed to urban areas prompted the government to encourage rural industrialisation as a means of absorbing surplus labour while also implementing draconian policies to stem citywards migration and promoting the growth of small (rural) towns. As the official slogan has it, 'leave the land but not the countryside; enter the factory, but not the city' (Marton 1998: 47). To a significant extent (and notwithstanding variations between provinces), non-farm workers have remained resident in their natal villages.[2] Indeed, it might be argued that China is unique in the extent to which rapid industrialisation has been achieved without a parallel process of urbanisation (and the rural–urban displacement of population that such a process implies). But this, as Parish *et al.* contend, may be a short-lived phenomenon. Those individuals showing the greatest propensity to migrate are the young and this may herald a change in patterns of moving and staying (1995: 715).

Marton's study of the Lower Yangzi delta offers a detailed insight into the processes underway. Kushan, a county-level municipality 55 kilometres northwest of Shanghai, had a population in 1996 of 580,000. The growth of non-farm activities in the region has been nothing short of remarkable – as it has been in China as a whole. In 1979 agriculture accounted for 35 per cent of total output. Between 1979 and 1996 the annual growth rate of industrial output averaged 32.7 per cent and the share of agriculture declined from 35 per cent to just 6 per cent (Marton 1998: 6).[3] This rapid expansion in industrial output led to an equally striking structural change in the economy of Kunshan. But, and importantly, this change was not accompanied by any great change in the level of urbanisation. In other words, Kunshan's recent economic history has been one of industrial growth without urbanisation. As the popular local slogan had it: 'In every village fires stir, and everywhere is belching smoke.'

While the national government saw the need to promote rural industries as part of a wider policy to restrict city growth, it has been local government support and investment in TVEs which has secured the prize of rural industrialisation (see Marton 1998 and Wang and Hu 1999). Thus a national tendency is rooted in local structures and processes. This is evident in Weixing Chen's (1998) reference to 'village conglomerates', direct descendants, in his view, of the collective enterprises that formerly came under the control of the brigade in the pre-reform commune system. But whether rural industries in China can be seen as manifestations of local or central initiative is not entirely clear. Market localism and political localism are critical ingredients determining economic success but the state still has ultimate sanction and the power to intervene. 'The VC [Village Conglomerate] today is at once a grassroots-level societal institution under the influence of the state's guidance in planning and control and an autonomous profit-seeking community on the market' (Weixing Chen 1998: 90).

[1] These figures are subject to some doubt. See page 67.
[2] Guangxi, for example, shows much higher levels of out-migration.
[3] A similar transformation is recorded in Wang's study of three villages in Liaoning province in Northeast China (1997b: 235).

And thirdly, growing attention has been paid to the inequality-narrowing effects that rural industrialisation may have by diverting investment to rural areas, promoting employment and producing a more equitable spatial distribution of resources. Rural industrialisation is felt to be particularly helpful in providing jobs for the landless and land-poor in the countryside (Sandee and Rietveld 1994: 117). It is for this third and final reason that many NGOs support rural industrialisation as a poverty-alleviating strategy and as a less disruptive way of fostering economic growth in rural areas.

Given this range of factors, it is perhaps not surprising that rural industrialisation means different things to different people. Saith (1991) makes a distinction between definitions which take a locational approach (i.e., and apparently self-evidently, it is industry in rural areas) and those which emphasise rural industries' developmental linkages with rural areas. Other scholars characterise rural industrialisation as the establishment, growth and development of industries in rural areas under local control. It is, therefore, locally based and at least in this sense endogenous, although external actors may be involved in the process of initiation. Because of the generally low level of skills and technology and limited funds for investment available to rural entrepreneurs these industries are also characteristically small-scale, low-technology and often craft-based.[2] Still other scholars would wish to draw the boundaries rather wider and include all and any industrial activities in rural areas from small-scale, village-based initiatives through to foreign-invested, capital-intensive enterprises producing for the export market. Parnwell offers the following definition: 'Rural industrialisation can be defined as a process involving the growth, development and modernisation of various forms of industrial production within the rural sector generally and rural villages specifically' (Parnwell 1990: 2).[3]

Until the late 1980s in many countries, rural industries and rural industrialisation were accorded a low priority. As a sector which was, in line with many definitions at least, characterised by low technology, low productivity and low returns, it appeared to offer little that would help promote higher incomes and brighter development prospects in rural areas (see Lanjouw 1999: 92–3). But from the late 1980s rural industrialisation became 'a fast-moving bandwagon' (Saith 1991: 459). In explaining its attractions to governments and academics alike, Saith highlights the failure of the industrial sector to generate jobs fast enough to absorb surplus rural labour. Furthermore, agriculture, he suggests, has reached the limits of its capacity to absorb labour in most countries in the Asian region, whether through extensification or intensification. With the industrial sector failing and agriculture transforming itself through the introduction of labour-saving mechanical innovations, rural industries provide one of the few avenues by which the employment and livelihood needs of rural people can be met. Or that, at least, is the theory.

---

[2] This is the form of rural industry which many NGOs seek to promote and support.
[3] In a later paper Parnwell makes it clear that, for him at least, rural industrialisation must be village-based industrial development (1994: 30).

Rural industrialisation: pieceworkers, cottage manufacturers and village factories

In their longitudinal village study undertaken in Laguna in the Philippines (see page 77), Hayami *et al.* (1998) note the profound change in the composition of the village economy brought about following the establishment, between 1991 and 1995, of eight manufacturing enterprises in the local area (namely, a paper mill and seven metal craft enterprises). The authors classify these operations into two categories: 'cottage manufacturers' and 'village factories'. Production in the cottage manufacturing units is based on simple technology with little capital investment, and each enterprise employs between 10 and 20 workers. Village factories are larger-scale, employing between 30 and 70 workers and are more capital- and technology-intensive. However, in both instances urban-based companies, producing for the export market, subcontract part of production to these rural enterprises.

Although Hayami *et al.* describe those operations in Laguna employing less than 20 people as 'cottage manufacturers' this terminology is deceptive, implying as it does household-based production which draws its labour from the family. It is notable that even these cottage manufacturers drew 30 per cent of their labour from outside the village. Indeed, no fewer that 38 out of 155 workers were sourced from other provinces (1998: 144). These out-of-province workers were housed, on site, in simple barrack accommodation and in this sense the cottage manufacturers show clear associations with large-scale export-oriented factories.[4]

The growth of rural industries, and the effect they can have on the local labour market is clear from the experience of the brick industry in the Klang district of the province of Ayutthaya 100 kilometres north of Bangkok in Central Thailand (see Arghiros 1997a, 1997b, 1998 and Arghiros and Wathana Wongsekiarttirat 1996). The brick industry here was certainly present, albeit in a fairly modest guise, in the 1970s. At that time all the brickyards were non-mechanised and small-scale. But as communications improved during the 1970s and 1980s, bringing Klang into easy overland contact with booming Bangkok, so the industry began to expand. In 1990 there were more than 1,000 brick-yards in the area, including a number using mechanised production techniques. The brick industry had become, by that time, a significant employer of local people, and especially of the land-poor and landless (Figures 8.1 and 8.2). But by the mid-1990s a serious labour shortage began to stifle further expansion. This can be linked to a 1993 decision by the Thai Board of Investment, as part of the government's programme to decentralise economic activity away from Bangkok, to nominate the area as one deserving generous benefits to entice large-scale industrial enterprises to establish operations in Ayutthaya. The result was a significant growth in factories producing textiles, footwear and consumer electronics for the export market. Because these jobs were usually better paid and certainly less dirty and arduous than working in the brickyards

---

[4] The authors describe the living conditions as 'rather awful' (1998: 144).

Rural industry and farmers in the city

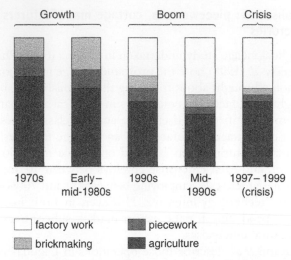

Figure 8.1   Local households: patterns of work in 'Klang', Central Thailand (1970s–1999)

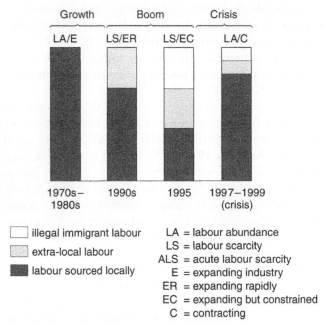

Figure 8.2   Make-up of the labour force in the brick-making industry, 'Klang', Central Thailand (1970s–1999)

of Klang, the new industries competed successfully for labour. Even with a marked flow of migrant labour from other parts of Thailand, the brickyards found it hard to recruit. Some employed illegal labour from Burma (Myanmar) to make up the shortfall but this proved unsatisfactory because of police raids and associated corruption. Arghiros writes of the situation in 1995:

Box 8.2   From craft to commerce: the expansion and development of the triangular pillow industry in Thailand

Triangular pillows, or *mon khit*, have been produced in Northeastern Thailand for many years. Until the 1970s production was small-scale and geared to local markets. Only in the mid-1970s and early 1980s did the industry become reoriented towards commercial production for sale more widely in Thailand, and even abroad. The expansion in non-farm work associated with Thailand's economic boom of the late 1980s and early 1990s created a heightened demand for pillows, both for personal use and for donation to temples. There was also an expansion in tourist demand as the pillows became desirable handicrafts. The boom made it increasingly difficult (and increasingly unattractive) for households in the Northeast to continue producing such pillows for their own use. As the industry expanded so local merchants became involved in the financing and organisation of production. Established producing villages like Ban Si Than in Yasothon province could no longer keep up with demand. A geographical division of labour appeared as villages specialised in particular stages of production. Ban Si Than, for example, concentrated on sewing and finishing (thereby maintaining their grip on the more remunerative parts of the production cycle), while weaving and stuffing was subcontracted out to neighbouring villages. This geographical and structural reorientation of production was encouraged and propelled by merchants (village entrepreneurs) who also played a central role in linking village producers with Bangkok-based wholesalers. As the relationship tightened and production escalated so the wholesalers, responding to consumers, began to feed design ideas back to the village. As Parnwell and Suranart Khamanarong argue, the recent history of the triangular pillow industry 'represents a classic illustration of a cottage industry which has managed successfully to adapt to changing market conditions by quite fundamentally transforming its system of production' (1996: 175).

Sources: Ikemoto 1996; Parnwell and Suranart Khamanarong 1996.

> **Most of the brickyards could produce at only half their capacity despite a buoyant demand. It is highly ironic that the very economic prosperity that gave rise to the brick industry is now strangling it by starving it of workers (1997a: 9).**

During this period of rapid economic growth, the tensions were not just between brickyards and other industries in the local area as they competed for scarce labour. The poaching of labour from competing yards became commonplace and owners also found that relations with local farmers were becoming increasingly strained. Owners would raise wages during the harvest season to keep workers from leaving the yards and, as a result, farmers had to try to recruit expensive agricultural labour from beyond the local area. In 1990 average wages in the industry were 50–60 baht; in 1996 these had doubled to 100–120 baht. Arghiros describes how he accompanied a brickyard owner on a fruitless journey through Thailand's poor Northeastern region in search of workers. Despite driving more than 1,000 kilometres and stopping at numerous settlements the owner was unable to recruit a single worker. Alternative occupations were simply more enticing (Arghiros 1998: 10). As in Hayami *et al.*'s study, Arghiros is at pains to emphasise both the speed and depth of change in Ayutthaya province, of which Klang is a part. Between 1989 and 1997, the gross provincial product of agriculture declined by 14 per cent while the gross provincial product of industry expanded 15 fold (1998: 7).[5]

The economic crisis which engulfed Thailand in mid-1997 presented the brick industry with another, even more serious, problem: lack of demand. The price of bricks halved, and then halved again between 1996 and 1998 (from 0.75 baht each to 0.17 baht). Most workers, both national and immigrant, were laid off as enterprises either shut down or contracted. The galvanised tin shacks which once housed the brickyards' non-local workers, and which were emblematic of the vitality of the industry during its heyday, lay empty.

Case studies such as these from the Philippines and Thailand emphasise that small-scale rural industrial enterprises are not, by their size, technology, orientation or ownership any more sensitive to local needs and conditions than larger-scale units. They may employ non-local people, sometimes in very poor conditions; they may distort the local labour market and undermine agriculture; and they are often tightly linked into the national and international economies and just as vulnerable to events beyond the local area. In other words, while some rural enterprises may, indeed, be inequality-narrowing and complementary to agriculture, these are by no means necessary characteristics.

Factories in the fields

Not only has there been an expansion in small-scale capitalist industrial enterprises, cottage and craft-based manufacturing, and piecework in the countryside (the latter is discussed in more detail below); rural areas have also been infiltrated by large-scale, capital and technology-intensive factories, significant numbers of which are foreign-invested and export-oriented. It seems that these factories offer the most enticing prospects for young people in rural areas. Arghiros (1998: 9), for example, remarks that while the brickyards of Klang may have competed successfully with agriculture for labour, they found themselves outcompeted by the export-oriented factories which government inducements were enticing into the area:

> It is important to appreciate that the most sophisticated of these modern factories offer workers non-economic rewards that brickyard labour will never provide. Workers identify work in such factories as 'modern' (*than samay*), and as an opportunity to surpass rural communities: women often go to work wearing make-up, young [men] use their regular incomes ... to buy motorbikes on hire purchase. To work in a modern factory reflects well on workers' self-identity (Arghiros 1998: 25).[6]

In Java, Wolf describes 'ten large-scale "modern" factories, driven by Western machinery and technology ... [squatting] in the middle of the agricultural land of two villages that still have neither running water nor electricity....

[5] This rate of change is comparable with the figures quoted in Box 8.1 for the Lower Yangzi delta.
[6] See the lengthy discussion of the attractions of modernity to rural youth in Chapter 4.

Plate 16 Making artificial fruit and flowers, Ban Hua Rin, Northern Thailand. These products are made for a Bangkok-based firm that exports most of its production. Inputs, bar labour, are sourced extralocally and workers are paid on a piecework basis.

Some . . . [nesting] in rice fields, disrupting neat rows of rice shoots with metal fences and guards' (1992: 109). Chantana Banpasirichote (1993) describes a very similar situation in rural areas of Chachoengsao in Central Thailand where factories are drawing people off the fields, and especially the young. 'The villagers see the arrangement [of having factories within commuting distance of the village] as an ideal solution, and the establishment of factories closer to the village has eased the families' decisions about sending their young to work in the factories' (Chantana Banpasirichote 1993: 33).

Class and the returns to work

While it is the low returns from agriculture which have propelled many rural people into non-farm work, this is not to say that non-farm work is necessarily more remunerative than farm work. This, it seems, is particularly true of piece-work. The finishing of peasant hats (or *ngop*) on a piecework basis in Arghiros' study of Central Thailand is so poorly paid that 'You can't do it fast enough to eat' (1997a: 6, see also Arghiros and Wathana Wongsekiarttirat 1996: 131). The same is true of basket-making in the same area and other forms of petty commodity production. Hayami *et al.* (1998) in the Philippines note the poor wages paid to the cottage manufacturing workers they studied in Laguna. Here the hourly wage was lower than that for agricultural work (although the daily wage is roughly the same: the farming day is usually shorter), a conclusion which mirrors that of White's earlier work in Java (1979 and 1993).[7] Cederroth (1995), in his study of Bantur in East Java, likewise notes the extreme marginality of much of the non-farm work undertaken in the village (pp. 111–16). Home-based basket-making offers a lower return than agricultural labouring, just 450 rupiah a day as against 1,000 rupiah (1995: 121).

Leones and Feldman's (1998) survey of 140 households in a resource-poor village in Northern Leyte, in the Philippines, illustrates the need for a nuanced and flexible perspective on non-farm working opportunities as avenues for development and upward mobility. The village in question is not only resource-poor, but transport links are also limited, precluding daily commuting to local centres for work. Nonetheless, the village has a high rate of out-migration (circulation). The authors note that the benefits to low-income families of non-farm work are closely related to the type of work available. Work with easy access and involving minimal capital inputs may narrow inequalities slightly, but activities with the highest returns require considerable investment, restricting access to higher-income families. 'In resource-poor villages, such as this one in Leyte, Philippines, where opportunities for earning income in agriculture are limited and where the cost of placement fees and transportation to obtain high-paying jobs are high, remittances can reinforce and exacerbate existing income differences' (1998: 802).

It has been convincingly shown that in many cases *in situ* non-farm work does not generate sufficient returns to support a household and certainly not sufficient to accumulate savings over time (see Saith 1991: 476). In Wolpe's (1972) terms, wages can be set at less than the cost of reproduction – creating a context where 'super-exploitation' can occur. However, the important point is that it is the articulation of capitalist production with peasant-based livelihoods which permits wages to be set at this level '. . . since in determining the level of wages necessary for the subsistence of the . . . worker and his family, account is taken of the fact that the family is supported, to some extent, from the product of agriculture . . .' (Wolpe 1972: 434).[8] In the case of Southeast Asia it is, furthermore, also usually the case that household subsistence is sustained by a combination of farm production and non-farm wages from more than one household member.

This is not to argue that rural-based non-farm work is necessarily poorly paid. However, it does challenge the assumption, purveyed by at least some alternative developmentalists, that rural industrialisation is a means to achieve greater equity without plucking people from their rural roots. Moreover, it seems that the occupations which pay best are those where rural people are employed in modern, large-scale, capital-intensive industries and not in the small-scale (or 'appropriate', in the development vernacular), craft-based operations which

[7] However, Hayami *et al.* note that people tend to prefer working in metal workshops because the work is less arduous and dirty than farming and because it offers year-round employment. Once again there are important cultural factors, and issues of preference, which condition patterns of employment.

[8] It should be noted that Wolpe was referring to South Africa and the employment of migrant workers from the Reserves in the Republic's mining sector. Nonetheless, the essence of the argument remains pertinent whether work is *in situ* or *ex situ*.

tend to attract the greatest attention. In 1911 Pleyte admonished supporters of small-scale rural industry in Java, writing:

> ... small-industry propagandists, you know only the outward appearances and judge native industries by them. Go into the kampungs [villages], visit the slums and hovels where the pieceworkers live and work, and see with your own eyes how the dispossessed in native society do their work and eke out their existence! (Pleyte 1911: 38, quoted in and translated from the Dutch by White 1991: 49).

For some rural households, having access to rural-based non-farm work enables them to remain in the countryside. Returns may not be enough to sustain a reasonable livelihood, but they are sufficient to keep the 'peasants' heads above water', to paraphrase Tawney (again) (1932).[9] It is this which, as several scholars have noted for Java, explains the remarkable resilience and persistence of small farmers in the face of the land-concentrating effects of technologically driven agricultural modernisation (see, for example, White and Wiradi 1989, Hart 1994, Maurer 1991, Cederroth 1995). However, while it may allow such households to persist and survive in rural areas, it does not allow them to prosper. In the past it might have been possible for a conscientious landless wage labourer to save enough money to buy a piece of land and become an own-account farmer. But the price of land today is such that for many this is impossible to contemplate – even should they wish to embed themselves long-term in agriculture (an increasingly unlikely prospect) in the first place.

This point has been highlighted in work from other areas of the developing world. In West Bengal, Chandrasekhar (1993) argues that men move into non-farm work when the harvest is poor. This, in turn, creates opportunities in the agricultural sector for women, who fill the 'slots' vacated by men. In this case, non-farm work is not indicative of any sort of economic dynamism but reflects seasonal crises in agriculture. Thus, it is perturbations in agriculture which explain patterns of non-farm work, rather than the other way around.

This brief examination of non-farm work raises questions concerning some of the discussion in Chapter 6 as to how individuals and the household should be conceptualised in the context of fast-changing rural areas (see page 83). The integration of farm and non-farm can only be fully appreciated when viewed from the perspective of the household or family, rather than the individual. Notwithstanding conflicts between different household members, and the multiple livelihood 'strategies' that may co-exist within a single family unit, this does not negate the utility and usefulness of the household as an organising unit. Indeed, extracting individuals from the household may make understanding their actions impossible.

Linking rural and urban industries

Hayami et al.'s (1998) and Tomosugi's (1995) work emphasises the tight links that are beginning to emerge between rural- and urban-based manufacturing

---

[9] 'There are districts [in Kansu, China] in which the position of the rural population is that of a man standing permanently up to the neck in water, so that even a ripple is sufficient to drown him' (1932: 77).

enterprises. In these cases, and for cost reasons, part of the production process has been subcontracted out to rural-based operations. In Hayami *et al.*'s Philippine case study these are small manufacturing enterprises, while Tomosugi's Central Thai case study focuses on individual householders employed on a piecework basis. In both instances, though, the products are finished in the capital city and then exported. Among China's Township and Village Enterprises (TVEs) relationships such as these are common. The TVEs provide the cheap labour and undertake the basic manufacturing operations while urban-based enterprises supply the technical, financial and marketing expertise and, in some cases, the raw materials (see Wang 1997a: 10). Wang in his study of Liaoning province – and as an interesting parallel to the emergence of the weekend farmer (who works in the city but returns at the week's end to work on the farm) – writes of the 'weekend technician' who is based in an urban area but carries out consultancy work for rural enterprises at the weekend (1997a: 11).

Tomosugi views the integration of the village of Tonyang into the global economy as having profoundly changed the relationships that exist between people, places and activities. In his view, the prosperity of Bangkok is increasingly supported and sustained by rural-based manufacturing, just as rural livelihoods are becoming increasingly dependent on non-farm work. Furthermore, the division of labour within manufacturing is becoming more overtly spatial, particularly with the gradual erosion of the rural–urban divide. Finally, the consumption practices which hitherto have been viewed as emblematic of urban lifestyles, have become part of rural life too (1995: 102–4). Again this has a nice parallel in Wang's study of China, where he writes, with tongue only partially in cheek, of the 'karaoke gap' in which the only difference between village and town is that 'we [villagers] cannot dance and sing *karaoke* as well as them [urban residents]' (Wang 1997a: 16 drawing on Guldin 1994).

Flows of labour between rural and urban areas also emphasise the fluidity of the situation. Rotgé (1992) notes the extent to which rural-based manufacturing in the vicinity of Yogyakarta (Java) is enticing workers from urban settlements into rural areas. Improving communications and cheaper land have encouraged such enterprises to locate in the peri-urban area, the extent of which is getting wider as cumulative improvements in transport make their effects felt. At the same time, the self-same improvements are opening up the possibility that rural people can commute to work in urban-based operations. This is reflected in Antlöv and Svensson's (1991) study of textile production in Majalaya, 35 kilometres southeast of the West Java city of Bandung. They describe how the bulk of labour for the textile industry in Majalaya is drawn from local villages where workers commute to their factories on a daily basis (Antlöv and Svensson 1991: 124).

A question which remains to be adequately addressed is how workers in rural industries should be categorised in class terms. Are they peasants or workers? Some scholars have challenged the assumption, so prevalent in studies of Southeast Asia, that the 'peasantry' dominate, in terms of weight of numbers at least. Ungpakorn, for example, maintains for Thailand that the 'working class

Table 8.1   Classes in Thailand: from peasant to worker (1996)

	Number (millions)	% of adult working population
Capitalists and the new middle class	0.7	2.2
Petty traders and family workers	4.0	13.5
Peasants	13.8	43.3
Workers	13.0	41.0

Source: Ungpakorn 1999: 5

is rapidly becoming the largest class in Thai society' (1999: 4) and provides figures for 1996 indicating that, in class terms, the number of 'peasants' and 'workers' are about the same at a little over 40 per cent of the adult workforce (Table 8.1). While there is little doubt that the balance between peasants and workers has fundamentally shifted in the last three to four decades, and will continue to do so – economic crisis notwithstanding – the task of allotting individuals to discrete classes is rather more difficult. As this chapter shows, the spatial interpenetration of rural and urban, and agriculture and industry; the difficulty of making a clear distinction between proto-industrialisation, petty commodity production, petty capitalist production and capitalist production; and the degree to which rural people (and especially rural households) flit between peasant-based and worker-based activities as they juggle multiple livelihoods pose a considerable conceptual and methodological challenge.

A virtuous cycle of farm – non–farm and rural–urban relations?

One of the key attractions of rural industrialisation is the belief that it complements traditional rural pursuits and, in particular, agriculture. This enables villages to maintain their economic viability and, importantly, their social coherence. Thus Funahashi (1996: 108) notes that the few villages in Yasothon province in Northeastern Thailand which have managed to keep their young men and women from drifting away to urban areas are those with a successful rural industrial base producing, for example, gemstones, triangular pillows or water jars. The question to be addressed in this section, at least in its stating if not in its answering, is comparatively simple: do farm–non-farm and rural–urban relations work out to the detriment or to the advantage of agriculture and agricultural production? Bray, in her study of the rice economies of Asia hazards that:

> intensive rice-farming dovetails very neatly with petty commodity production, which requires very little capital to set up a family enterprise, and absorbs surplus labour without depriving the farm of workers at times of peak demand. It can be expanded, diversified or contracted to meet market demands, but the combination with the rice farm guarantees the family's subsistence (1986: 135).

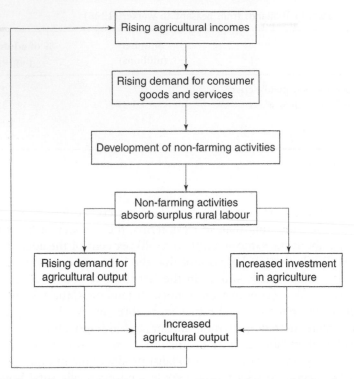

Figure 8.3 Agricultural production, innovation and non-farm rural enterprises
Source: Rigg 1997

In their work on extended metropolitan regions, scholars like McGee and Ginsburg have suggested something rather similar when they highlight the manner in which rice agriculture and non-farm work dovetail, both spatially and in terms of the division of labour (see page 59). Furthermore, these integrative urban-cum-rural regions are seen to have a certain level of stability and persistence.

The importance of the question as to whether farm–non-farm relations are complementary or conflicting stems from a range of work, initially rooted in Asia but since applied to other parts of the developing world, which postulates that there is a 'virtuous cycle' of farm–non-farm (and sometimes, rural–urban) relations. In this sequence, rising agricultural incomes create a demand for consumer goods and services. This spurs the development of non-farm activities such as motorcycle and watch repair shops, which help to absorb surplus farm labour. This, in turn, further boosts demand for farm output while contributing cash for investment in agriculture, thus both stimulating and permitting yet additional increases in agricultural production (Evans 1992: 641, see also Evans and Ngau 1991, Nipon Poapongsakorn 1994, Effendi and Manning 1994: 230, Koppel and James 1994: 290, Lanjouw 1999: 95, Meagher and Mustapha 1997) (Figure 8.3).

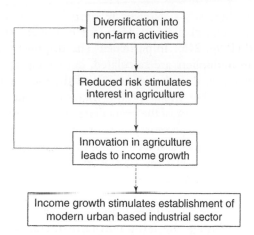

Figure 8.4 Agricultural production, innovation and urban enterprises

Grabowski (1995) and Haggblade *et al.* (1989) similarly see non-farm work as a critical component in generating innovation and productivity increases in agriculture. Grabowski suggests that because such activities lead to a diversification of income sources they reduce risk and this creates the incentive for innovation on the farm (1995: 53). On this basis, he contends that rural-based non-agricultural activities should be accorded infant industry status. The result of the neglect of such protection:

> . . . is that innovation in agricultural production becomes more difficult to introduce, domestic markets grow very slowly, and modern urban-based industry (increasing returns) finds it extremely difficult to survive in the Third World without continuous protection. Thus, urban-based manufacturing never grows up and never finds itself able to compete internationally (Grabowski 1995: 53–4).

Haggblade *et al.* (1989) take a slightly different line, arguing that agricultural growth leads to greater consumption expenditure and this provides a stimulus for the expansion of non-farm activity. Not only does such rural non-agricultural work spur agricultural innovation, but Grabowski (1995) also believes that the development of a vibrant non-farm sector is critical to urban industrial development. For, in his view, the emergence of a modern industrial sector in urban areas is contingent on rural incomes rising sufficiently to create a domestic market large enough to support such a sector (Figure 8.4):

> Innovative activity by farmers is inhibited by a lack of inputs, the limited development of the market for agricultural goods, and the risk attached to such activity. The spread of nonagricultural activities in rural areas allows farm families to diversify their sources of income and reduce risk, and this predisposes farmers to choose more innovative activities [in agriculture]. The resulting income growth stimulates further growth in cottage manufacturing and nonagricultural activities, which in turn stimulates further agricultural growth, eventually resulting in the establishment of modern industry as domestic markets expand dramatically (Grabowski 1995: 50).

This notion that there is a synergistic relationship between farm and non-farm has been vigorously challenged. Hart, for example, regards such a view as 'deeply suspect' (1996: 246). In particular, she disputes the basis on which the farm–non-farm multipliers are calculated, contesting the assumption that agricultural growth will lead to the expansion of the local non-farm economy. Instead, she postulates that much of the increase in income generated by agricultural growth will leak out of the immediate vicinity into the wider national economy (1996: 251–2). Furthermore, there are cases where rural industrialisation has challenged the agricultural sector and, rather than stimulating increases in productivity, it is causing just the reverse effect. This is true, for example, of Chantana Banpasirichote's study of Central Thailand, where the process of occupational shift in the countryside is directly leading to labour shortages and thereby to a decline in agricultural productivity (1993: 33). So, the notion that rural households can bridge the farm–non-farm divide with ease overlooks many of the issues and processes highlighted in the previous chapter. To begin with, it ignores the social and cultural changes which have made farming, for many people, an unattractive proposition. Second, it overlooks the conflicts that can arise between men and women and between the generations. Third, it disregards the environmental costs of rural industrialisation (and some alternative agricultural practices, like prawn farming). And fourth, it tends to ignore or play down the political economy of rural change.

So, if it is not a buoyant local agriculture, what is promoting the evident rural industrialisation of Southeast Asia? One possibility is that offered by Tambunan in his study of West Java (1995) in which he identifies the growth of rural industries as a response, in large measure, to rural poverty.[10] Many people in the nine villages of Ciomas district where he undertook his fieldwork embraced such activities as a 'last resort'. In noting this, Tambunan links his study with work on the role of the rural non-farm sector more generally, where scholars have noted the prevalence of distress diversification (see page 88).

An alternative approach is to look beyond the rural in search of an explanation. For it would seem that the growth of industries in rural areas lies partially in the strategies of urban-based industrialists searching for greenfield sites away from the congested towns and cities (Hart 1996: 253). This, in turn, is linked to the policies of governments and in particular their efforts to achieve a more equitable distribution of the space economy by promoting regional development and the industrialisation of lagging or peripheral areas. But, at a more profound level, it is also questionable whether – and this resonates with Hoggart's work (1990) on the developed world – the 'rural' is a very useful category when it comes to interrogating the process of rural industrialisation in Asia. As the case studies noted in this chapter demonstrate, there is no obvious division or distinction between rural and urban industrialisation, nor between the workers who labour in such enterprises.

---

[10] This mirrors the conclusions of Zoomers and Kleinpenning's (1996) work in Central Paraguay where they see the role of Asunción not as a catalyst for agricultural growth (as described earlier in this section) but as a 'safety valve' for the rural poor. See also Chandrasekhar on West Bengal (1993).

However, this is not to say that rural non-farm development has no links with agricultural development. There are case studies where rural industrialisation is apparently closely associated with the surpluses generated by the local farm sector. Moreover, 'the existing literature', Lanjouw contends, 'points to a potentially strong relationship between the rural nonagricultural sector and rural poverty' (1999: 95).

Farmers in the city

There is little doubt, and notwithstanding the statistical deficiencies noted in Chapter 5, that the decades since 1945 have seen a substantial influx of rural people to urban areas. In some instances there has been a wholesale displacement of families to urban centres. Their links with rural areas have been severed, and their futures have become dependent upon opportunities available in the towns or cities where they live. As Hobsbawm writes, 'When the land empties the cities fill up' (Hobsbawm 1994: 293). But for the majority, such a clean break does not occur. As was noted in Chapter 6, families become split between urban and rural 'homes' and livelihoods. For some, their stay in town is short-lived. A temporary sojourn, perhaps with a seasonal character. For others it may be more permanent. But their rural links remain important, economically and culturally.

Bruner, writing of Toba Batak migrants in the city of Medan in North Sumatra in 1961 observed:

> The Indonesian village is often presented as an isolated world in itself, a closed self-contained entity, relatively static, while the city is the true dynamic centre ... But this conception does not apply to North Sumatra. The rural and urban Batak are linked through a complex communication network in which Western goods and ideas do flow from city to village, but the flow of people and of the moral support and vitality of the adat is primarily in the other direction. *The cultural premises and roots of urban Batak life are to be found in village society* (1961: 515 (emphases added)).

In writing this Bruner is suggesting that Toba Batak living in Medan may have been spatially displaced to an urban centre, but their social and cultural identity remains rooted in the countryside. Four decades on from Bruner's study and the situation is both simpler, and more complex. The idea of there being an identifiable 'village society' and (by implication) an 'urban society' too is difficult to sustain, except in an extremely blunt and generalised manner. Furthermore, at least in terms of economic activity, it is not so much the contradistinction of rural and urban which is of interest, but their blurring.

DiGregorio's study (1994) of scavengers in Hanoi provides an excellent example of the way in which farmers insinuate themselves into the urban fabric. The title of his study is instructive: *Urban harvest: recycling as a peasant industry in northern Vietnam.* As was noted in Chapter 6, the majority of those working in the recycling industry come from rural areas, mostly as temporary migrants working to supplement their rural incomes. They live in boarding

Plate 17 Fruit and vegetable canning, Chiang Mai, Northern Thailand. This firm, with links to a Hong Kong–based company, exports its production largely to Asia and Europe.

houses and maintain close ties with their villages of origin. Sicular, in his earlier study of scavenging in Bandung, Java, conceptualises scavenging as essentially a peasant form of production translated to the urban context. He makes this assertion on the basis of the social formation within which such work is structured and writes of 'pockets of peasants' in the urban setting (Sicular 1991: 154–5). Like farming, scavenging is based on a household unit of production. DiGregorio, while he rejects the notion that scavenging is another form of peasant production, does concur with Sicular to the extent that he highlights the tight links between scavenging and farming, and between town and country:

> The linkage of work in the recovery industry with rural production colors much of the relations of production, reproduction, and consumption in the community of waste materials recoverers. The most obvious of these connections is the leading role played by agricultural cycles in determining when and who is able to work in Hanoi (1994: 95, see also Figure 5.1 on p. 66).

Men, because of the gender division of labour in agriculture, are the first to be released from work in the fields. Women and children, who are heavily involved in transplanting, weeding and harvesting, tend to be released rather later and for a shorter period. The respective roles of men and women in the scavenging industry are thereby contingent on the demands set by rice cultivation. Their temporary presence in the city also explains why nine out of ten scavengers stay in boarding houses where they live, sleep and eat close to other rural people, usually from the same district. Sicular also sees the character – indeed the very presence – of the recycling industry in Bandung as being closely linked to conditions in the countryside and, in particular, the penetration of capital into rural areas and the social disruption and economic differentiation that this has produced (1991: 142). Thus, understanding the presence, structure and operation of the recycling industry must be rooted in an appreciation of the social and economic relations of production embracing farm and non-farm, local and extra-local. Just as *ngop* finishers in Central Thailand (see above) can be paid at a rate which is so low that 'you can't do it fast enough to eat', so returns to recycling are highly marginal for the same reason: the income generated is supplementary to farm-based production and payment can therefore be set at less than the cost of reproduction. How the links between farm and non-farm production are played out, and whether there is a tension between the two areas of work, is not fixed. In some cases it appears that it is to the detriment of agriculture (as noted above and in more detail in the previous chapter). In other instances, such as DiGregorio's case study, it would appear that non-farm work is moulded around the more important demands and exigencies of farming.[11]

---

[11] Wang's (1997a) study of Tuchengzi village in Liaoning province in Manchuria, Northeast China provides another example. Suitcases and leather bags have been produced in the village since the mid-1980s and make an increasingly important contribution to rural livelihoods. Nonetheless, the inhabitants were unwilling to give up farming and 'during the busiest farming seasons most of the women [continued to work] in the field during the day and made bags and suitcases at night' (1997a: 8).

Plate 18 A member of the Ban Hua Rin women's group making shirts for sale at the local community shop. This small workshop is based in the grounds of the village monastery and has the enthusiastic support of the head monk.

Reformulating the agriculture–industry divide: rural industry for rural areas?

Dualistic models are, for the moment at least, out of vogue. Formal–informal, developed–developing, South–North, all such twofold divisions have come under intense scrutiny, for the most part because they are perceived to oversimplify a complex world where nuance and multiplicity are the norm. Grabowski argues that the traditional division of the economy into the industrial and agricultural sectors distorts 'the development process and agriculture's role in it' and instead proposes a three-sector model where the third sector is non-agricultural production in rural areas (1995: 41). In his schema noted above, the rural non-agricultural sector plays a strategic role linking agriculture and industry. It creates the conditions for agricultural diversification and productivity increases and this, in turn, promotes the growth of urban-based industrial activities. Certainly, such a three-sector model helps in bringing more conceptual precision to the debate and, importantly, takes industry out of its traditional urban ghetto. But in a sense, and as I have argued elsewhere, it does not go far enough because it merely replaces a simplistic twofold model with an only slightly less simplistic threefold conceptualisation. The reality of rural–urban relations and activity would 'seem to call for a rather more subtle model of how different activities complement, cross-subsidize and inter-relate' (Rigg 1997: 268). Indeed, much of this book has been an attempt to expand on these very issues.

In trying to explain the infiltration and dispersal of industrial activity into rural areas the focus should be on 'the multiple, nonlinear, divergent trajectories through which industrial capital in numerous guises encounters and intersects with agrarian conditions constituted through particular local histories together with broader processes of agrarian transition, industrialisation, and urbanisation' (Hart 1996: 254, and see Hart 1997). While at one level we can say that the 'industrialisation of the countryside' is a theme with wide application – in Asia and beyond – this is not to say that the underpinning mechanisms are the same. Hart (1996: 254–9) compares, for example, the contrasting cases of Taiwan, coastal China, and Malaysia. In Taiwan, rural industrialisation emerged as a significant phenomenon from the mid-1970s and was 'endogenously' driven, with rural-based entrepreneurs who had gained their knowledge and experience in urban factories setting up small, family-owned enterprises on their land (Hart 1996: 255–6). In coastal China, rural industrial activity has been intimately associated with TVEs (see the discussion in Box 8.1) which have close associations with the local government apparatus, while in Malaysia, industrial expansion in, for example, the Muda area of Kedah has been driven by foreign investment. Furthermore, there are equally marked differences within individual locales in single countries. Thus areas of Malaysia, Indonesia, the Philippines and Thailand exhibit examples of craft-based manufacturing, petty commodity production, small-scale capitalist industrial enterprises, and large-scale and (foreign) capital-intensive factories, each with its own internal logic and trajectory. So, just as the historical experience of the agrarian transition in various parts of the world is marked by difference (see the discussion in Chapter 2), the same is true of the forces underpinning rural industrialisation. While we may be able to observe that rural industrialisation is a global trend, the specific processes that underpin and explain the trend are significantly different.

For proponents of rural industrialisation one of its attractions (along with those mentioned towards the beginning of this chapter) is the belief that it is especially geared to producing goods and services appropriate for rural populations and agricultural production. Kirkby and Zhao Xiaobin in their study of China lament the end of the commune era when 'rural industries provided substantial support to local agricultural production by supplying farm tools, fertilisers, machinery repair services, and the processing of foodstuffs . . .' (1999: 277). They apparently rebuke rural manufacturing enterprises for 'rush[ing] to develop the most lucrative lines, with little reference to local needs'. By the 1990s, TVEs in China's more developed coastal provinces could be distinguised 'hardly at all' from urban-based industry (1999: 277).

The experience of China, along with the case studies from the Philippines, Thailand and Indonesia noted above, illustrates the difficulty of pigeonholing rural industries, let alone separating them in any meaningful sense from urban industries. The difficulty confronting analyses of the role of rural industries is that the sector – if, indeed, it can be categorised as such – is becoming increasingly diffuse and heterogenous. There are, it seems, as many definitions of rural industry as there are examples of it. TVEs, as noted in Box 8.1, are at the cutting edge of China's industrialisation efforts. India's *khadi*[12] and village industries (KVIs), by contrast, are small-scale, traditional and locally oriented – and representative of the stagnant nature of much of the subcontinent's non-farm sector (Fisher *et al.* 1997: 8). Thus in China and India rural industrialisation has come to symbolise very different things.

Furthermore, definitions of rural industries have a habit of falling foul of the dynamic nature of the sector. At a country level this is again nowhere clearer than in China where Mao's backyard furnaces serving commune needs and using local ingenuity have been transformed into advanced industrial units producing for the national and international markets. But a similar trajectory can be identified in individual enterprises right across the Southeast Asian region. They have established links with urban-based industries and entrepreneurs, and in expanding their commitments in this way have begun the process of breaking their formerly close association with the local area. At the same time, rural areas are being infiltrated by new enterprises, sometimes foreign-invested and often with few rural linkages, except in so far as they employ rural labour. This is akin to the 'ruralisation' of industry in the West (see Marsden *et al.* 1990: 9) where capital has infiltrated the countryside in search of profit. Industry in rural areas is undeniably important. But how it is important and to whom, and how it came to be important are questions which can only be satisfactorily answered by reference to local political economies (and their histories) and how these articulate with the national and international.

---

[12] *Khadi* is the home-produced cloth eulogised by Mahatma Gandhi as representative of village self-reliance.

Chapter 9

Re-making the Southeast Asian countryside

Nothing in human life has changed more in ten centuries than the world of work. . . . The most obvious symbol of that is the shift from farming . . . (*The Economist*, 1999: 22).

Hybrid lives, hybrid work

Currently, much work on how rural areas and rural people in Southeast Asia are adapting to the new challenges and opportunities that confront them has emphasised such themes as integration, interpenetration, occupational multiplicity and hybridisation. In other words, rural households remain torn between rural and urban, and between agriculture and industry (with a considerable amount of overlap between the four). What is not clear is whether this emerging pattern will prove enduring. Is it just a stage en route to a more fundamental restructuring of rural areas and rural people's lives? Or is pluriactivity, as it is also known, an outcome of these processes and therefore likely to persist? This question, of course, is one with a long history because it links back to the debate over the survival of the family farm (see Chapter 2). The evidence from Southeast Asia, and from other regions of the world, points in both directions. This is to be expected.

In Europe pluriactivity is spreading as farming's ability to meet household needs has been eroded. What is significant, though, is that in classical terms these countries have, in the most part, successfully negotiated the agrarian transition. This would seem to support the view that the 'agrarian question' is never resolved; it is merely reformulated in the light of the challenges and tensions that economic and social change inevitably present (see the discussion in Chapter 2). Ploeg shows for the Netherlands that a minority of 'farm' households – in some areas just one-third – earn their incomes solely from farming. Agriculture, as he puts it, 'is increasingly [just one] part of a wider income strategy' (1993: 247, see also Jervell 1999, and Kalantaridis and Labrianidis 1999). Moreover, this is clearest among younger farmers. Thus recent changes in the developed Netherlands mirror, in several respects, those evident in developing Southeast Asia.

In her work on Africa, Bryceson (1997a) writes of a widespread *process* of 'de-agrarianisation' – rather than a *state* of pluriactivity. The implication seems to be that the ultimate outcome will be the extraction, in both economic and social terms, of the rural person from the countryside and from agriculture. This seems to be the case in Francis and Hoddinott's study of migration and differentiation in Western Kenya, where they note that 'the material links between [long-term migrants] and their places of origin are now weakening' (1993: 126). Migrant sons are remitting smaller and smaller shares of their income. The authors quote figures from studies undertaken in the 1970s and 1980s of 20 per cent of total income being remitted home; the share recorded in this late 1980s study is just 6 per cent and they find it 'hard to imagine their keeping their rural ties active in adulthood' (1993: 140).

This study from Kenya hints at a tension which was noted earlier: between household and self. In Southeast Asia it has become commonplace to refer to what might be termed the 'dutiful daughter syndrome' in explaining the tendency for young women (particularly) and men to continue remitting a portion of their income to their natal families in the countryside even after they are well established in urban work. This is sometimes justified because urban work is seen as just a stage in the life course. Therefore remitting money to keep their rural options open makes good long-term sense for these individuals. This is particularly true if they stand to inherit the farm, or a portion of it. But increasingly it is becoming clear that these migrants will never return home. Their best interests, and especially if they have accumulated human capital through education, are served by cutting themselves off from their agricultural roots and extracting themselves from the family and its associated household economy.

The experience of East Asia is notably different from that of both developing Africa and Southeast Asia, and developed Europe and provides support for the possible long-term persistance of pluriactivity. In Japan and Taiwan the 'large majority' of farm households earn their living from on-farm production coupled with off-farm work (Francks *et al.* 1999: 15). Furthermore, it is worth remembering that both countries have passed through an agrarian transition. Why should the experiences of Japan and Taiwan (and also South Korea, but to a lesser extent) be apparently so different from that of other developed countries?

Some scholars (for example Bray 1986) have suggested that paddy rice farming is not conducive to the sort of economies of scale possible in temperate farming. On this basis they make a case for the uniqueness of wet rice agriculture: rice farming will always be based on the small-scale family farm because it cannot be otherwise.[1] A second possible explanation is that massive government support for agriculture, clearest in Japan, where the insulation of the rice market became almost legendary, explains the persistence of the small farm. Without this support, small farmers – pluriactive or not – would have been forced off the land. A third possibility is that the land reforms pushed through in the aftermath of the Second World War, in particular in Taiwan and South Korea, stemmed any tendency towards land concentration, thereby providing small farmers with a breathing space which they are still enjoying. Finally, we should not ignore the ability of farmers to adapt to new opportunities and

---

[1] The presence of large-scale rice farms in the United States would seem to suggest otherwise.

increase their production of non-traditional crops to meet new economic opportunities and sustain livelihoods in the face of rising needs. In Taiwan, for example, Francks *et al.* note the shift of farmers out of rice and into higher return fruit, vegetable and livestock production (1999: 186).

But while there are reasons to argue for the uniqueness of the East Asian experience, it may be that, despite appearances, the region is not significantly different at all. It is just that because the agrarian transition has been collapsed into a few decades, rather than a few generations as it was in Europe, the generational changes which are so critical have yet to work their way out. Or to put it another way, we are still waiting for the original farmers to retire or die. Francks *et al.* note that by the early 1990s one-third of the agricultural labour force in Japan was over 65 years old (1999: 82). They continue: 'In the more remote and mountainous rural areas, the absence of successors has led to the abandonment of agricultural land, as household heads age and die, and to environmental problems, as irrigation infrastructure and field systems fall into disrepair' (1999: 82). East Asian governments appear to recognise the impending crisis in agriculture and envisage the creation of a cohort of professional, full-time farmers to fill the emerging rural void and provide for the countries' agricultural needs.

Surveying the literature from the developed and developing worlds, one thing becomes clear: there are multiple agrarian paths and, moreover, there is 'continuous revolution'. In Europe, pluriactivity is back on the agenda as farm households struggle to maintain their presence in agriculture in the face of falling returns. In East Asia, professionalisation and land concentration are the watchwords as pluriactive households struggle to maintain their incomes in the face of falling returns. The experiences of other parts of the world may offer insights into the possible, but they in no sense offer a model of the agrarian transition as it will emerge in Southeast Asia.

An Italian diversion

Nowhere is this clearer than in the historical experience of the Como and Brianza region north of Milan in Italy. This vignette of social and economic change, wonderfully recounted and analysed in a series of publications by Bull (1987 and 1989) and Bull and Corner (1993), resonates strongly with contemporary processes in Southeast Asia while also forcefully demonstrating the uniqueness of each agrarian transition.

In their book *From peasant to entrepreneur* (1993), Bull and Corner examine the transition from an agricultural to an industrial economy in the Como and Brianza region.[2] But their interest is not in the process of structural change per se, but rather in the way in which peasant families have adjusted to transition. They note that the essence of change in this area of Italy from the early nineteenth century was that families remained enmeshed in agriculture while at the same time becoming increasingly dependent on non-agricultural work,

---

[2] This section draws largely on Bull and Corner's (1993) book supplemented by Bull 1987 and 1989.

a situation of pluriactivity. Moreover, this was not a temporary, short-term phenomenon – a passing phase as peasants became workers – but existed for 60–70 years. Importantly, peasants turned their attention to non-farm work '*in order to be able to survive on the land*' (1993: 3 (emphasis in original)). They point out that while considerable scholarly attention has been paid to the formation of an urban working class in Italy (and particularly in the north), the formation of a peasant-worker or worker-peasant class has been underplayed. For some considerable time, they say, 'production in manufacturing was realised more by workers who were not working class, in the conventional sense of the urbanised industrial worker, than by those who were' (1993: 4). The authors are interested, particularly, in the important question of why peasants did not leave the land.

The evolution of the peasant-worker family unit proved to be a particularly flexible and adaptable means of exploiting opportunities in an unpredictable world. In the 1820s and 1830s the price of silk rose by more than 100 per cent and the Lombardy-Venetia region became Europe's major producer. But even in such a buoyant market, farmers did not turn their land and their labour over to silk production. Instead 'silk was grafted on to the existing [wheat-based] system' (1993: 18). By the middle of the nineteenth century peasants had begun to leave the land during the agricultural slack season to work in the expanding factory sector largely geared to the silk industry (in particular silk winding and throwing mills). The expansion of this sector in the upland valleys of the region was intimately linked to the availability of cheap labour. Most of this labour was supplied by unmarried women and children who were not only malleable and docile but could also be paid at less than the cost of their reproduction.[3] Furthermore, during slack periods or downturns in the industry they could simply be laid off, to be re-employed when conditions improved. They were both tied to the land, in the sense that agriculture provided for their basic (very basic) subsistence but were also free to labour off the land should work be available. The outcome was that the 'industrialisation of the hill regions was realised . . . without the expulsion of labour from agriculture' (1993: 27). By all accounts conditions in the factories were horrendous, even by the standards of the day. Yet the meagre incomes which women and children brought back to their husbands and fathers on the farms allowed sub-livelihood farms to survive. Before long agriculture and industry had become complementary activities – not alternatives – but nonetheless complementary activities where agriculture remained dominant (1993: 29).[4] Traditional family structures, and in particular the rigid patriarchal system of the area remained surprisingly resilient, despite the involvement of women and children in industrial work. Men – who

---

[3] Bull (Bull 1987: 110–11) disputes the assumption that women tended to leave the factories once they got married and, more particularly, when they started to have children (as commonly reported in modern factories in Asia). She notes the dramatic increase in the number of orphans and also the presence, in some silk spinning factories, of nurseries and feeding rooms. What she does not dispute, however, is that by the age of 30, when women might have had two of their average five children, most would have left the factories and returned to their husbands and the land.

[4] The authors note that for a long time women and children would continue to desert their factories when labour was needed on the farm (Bull and Corner 1993: 30).

remained peasants even while their wives and children were becoming workers (or worker-peasants) – dictated patterns of accepted behaviour, which perhaps explains why factory work (in contradistinction to Southeast Asia) was regarded as demeaning:

> Whereas the female members sometimes shared the labour of the male members (at the time of cocoon raising for example), the opposite never took place. The head of the family was a full-time farmer and, though he was a sharecropper, he viewed the land he cultivated as his own and clearly aspired to the status of landowner. The fact that his daughters – possibly even his wife – worked in a silk factory was considered a dire necessity, as their wages were a means to preserve the family as a rural family or, to put it more precisely, to preserve his status as a farmer and possible future landowner (Bull 1987: 112).

Young women were caught in a trap where a rural/agricultural patriarchal culture dominated their lives even while they were working as factory labourers. The fact that these factories were located in rural areas kept women workers at home, preventing their escape. The risks to unmarried women of simply leaving home for the town were great: they usually knew no one there and, in any case, jobs for silk workers were situated in the countryside where the factories were located. In this way, women and children, Bull concludes, were exploited at three levels: by landowners, due to their position as members of the rural family; by male household heads, due to the prevailing patriarchal system; and by factory-owners, as industrial workers (1987: 117).

This pattern of pluriactivity lasted through to the 1890s, when there is evidence that families were increasingly turning their attention to industry. But, Bull and Corner quickly add, these families still had 'no intention of leaving the land'. Even while sub-division of holdings was continuing apace, and incomes from factory work were rising (especially from 1900 onwards) families kept their toeholds on the land. A farm, however small, and a farmhouse were guarantees against hard times – an insurance policy. But the diminishing size of land holdings did force farmers to reallocate time and effort increasingly towards industrial work. 'Why', Bull and Corner muse, 'did peasant families, for whom agriculture had for so long been synonymous with poverty, not abandon the land when other opportunities appeared? And why did peasant families stay together, despite centrifugal pressures?' (1993: 49). One answer, the authors conclude, is that rural hardship was preferable to urban poverty. But more important was the fact that families operated as units. It is not possible to understand individual action except in terms of family action. Families had mixed interests and these could only be met in terms of a mixed family economy.

This worker-peasant family model continued to have its own internal logic through to the mid-1920s. At that point families began to abandon the essentially defensive (i.e. risk-minimising) strategies of the past for livelihood strategies which would allow a measure of accumulation (Bull and Corner 1993: 77). Men as well as women became involved in industrial work and, furthermore, in industrial work beyond the silk industry. There was also significant expansion in small-scale industries as some worker-peasants became entrepreneurs. From

Plate 19   A grand mausoleum in the ricefields of the Christianised Batak of Lake Toba, North Sumatra, Indonesia. Edifices such as this reflect the fortunes that have been made by Batak who have left home.

the 1930s the peasant-worker was in terminal decline and in the 1951 census a majority of households gave a non-agricultural activity as their main occupation (1993: 100).

What's in it for rural people?

The Como-Bianza experience is not one that helps sustain the view that non-farm work is good for you. Perhaps it is the European historical legacy exemplified in this vignette of change in Italy which explains why an assumption permeates much of the literature that the processes outlined in this book, and sometimes encapsulated in the term 'de-agrarianisation', are negative ones. Certainly, there are developments which it would be hard to gloss as positive. The undermining of the environmental integrity of long-established agricultural areas and the social ruptures that can go hand in hand with spatial dislocation, for example. But, and in the round, the processes of rural change outlined here are broadly positive and developmental for the majority of rural inhabitants.

The debate over what is good and bad for the countryside appears at its most muddled when it comes to assessing the effects of rural industrialisation. There is a deep-seated view that local work is better than extra-local work. Keeping people 'on the farm' is less destructive and more developmental than seeing them drift to towns and cities. Furthermore, the type of rural industry is also seen as important in determining whether the developmental outcomes are likely to be positive. Small-scale, locally owned and managed industries which source their inputs from the countryside and serve the local community are commonly viewed as defining features of desirable rural industries. The intrusion of global capital into rural areas and the expansion of industrial activities which have few linkages with the rural economy are seen, in some quarters, as anti-developmental.

When processes of rural industrialisation are examined in detail, it is not at all clear why any of the above assumptions should apply. Why should migration to work in cities be indicative of a loss of human resources, and essentially undermining and destructive of rural communities, while employment in a village-based craft industry, however poor the returns, is viewed as positive in developmental terms and symbolic of rural people's ingenuity and self-reliance?[5] As the discussion in Chapter 8 demonstrated, many traditional rural industries are poorly paid, provide little chance for upward mobility and skill acquisition, and do little to lift people out of poverty. It is modern industries, sometimes foreign-invested and often with few linkages with the local area beyond employment, which allow people to cash in their carefully and sometimes painfully acquired human capital and attain a degree of upward mobility. Francis and Hoddinott, in their study of Kenya noted above, argue that 'the most successful individuals are those people who have effectively cut their links to the rural economy, or who have obtained locally based wage employment . . .' (1993: 142). This viewpoint on the African experience is supported by Bryceson, who writes that de-agrarianisation is not only inevitable, but a blessing rather than a blight, which 'harbours . . . the possibility of unleashing the largely latent creative potential of a continent' (1997b: 255). The talent of Southeast Asia's rural population has, to a significant degree, already been unleashed and, furthermore, this has been to the benefit of the majority who have engaged in such work.

Blurring the divide

An important element in the deep-seated antipathy towards both rural people taking up non-local employment and the infiltration of capitalist enterprises into rural areas is the tendency towards economic (agricultural–industry) and spatial (rural–urban) partitioning. This, in turn, is associated with the rather wider debate over whether there is an urban bias in development. In this view of things, and to simplify, urban areas and industrial activities are bad for rural areas and agriculture.

Through their actions, households have eroded both these divisions – sectoral and spatial. In this way 'terms of trade have been "internalised" within the extended family, straddling the rural-urban physical gap' (Jamal and Weeks

---

[5] Bryceson (1996) notes in her discussion of de-agrarianisation in Africa, the prevailing tendency to treat spatially distinct activities as somehow qualitatively different in terms of their effects on rural areas and argues that such a view makes little sense.

1993: 124).[6] I have written before (see Rigg 1997: 276–8 and 1998a) on the difficulty of applying notions of urban bias to a rural context which is increasingly fluid and interpenetrated: 'The difficulty is that conceptual structures and semantic divisions force scholars to prise people into categories which are no longer . . . valid' (1997: 276).[7] This blurring of the rural and urban, driven by the processes of interpenetration discussed at several points in this book, is not peculiar to Southeast Asia, or even to Asia, but has been noted in studies from across the developing world. It is also a common theme in work on the developed world. In his study of return migration and counter-urbanisation in Andalucía, for example, Hoggart concludes by stressing the 'turbulence' of rural populations (Hoggart 1997: 147).

As a result of this blurring of spatial and sectoral categories, the countryside is becoming a mixed-interest space. And it is doing so in two principal ways. To begin with, rural people (both at an indivdual and a household level) have a presence in several sectors and spaces simultaneously. This means that defining rural people as a 'class' with interests distinct from urban people (or agriculturalists as distinct from non-agriculturalists) is becoming an impossibility. Ritchie notes this in his study of northern Thailand where 'As boundaries between rural and urban blur, it is becoming more and more difficult to speak of a distinct "peasant" society' (Ritchie 1996b: 2).

But equally importantly, and perhaps less self-evidently, rural spaces themselves are being infiltrated by new interest groups and activities. Writing of rural areas in the developed world, Marsden et al. state: 'The decline of the postwar certainties, most notably agricultural productivism and its corporatist structures, has opened the way for the emergence of a more differentiated countryside, one whose trajectory is no longer determined to the same degree by the fortunes of a single industry but by a much more complex assemblage of economic, social and political elements' (1993: 185). This perspective has relevance here, too. The scope and level of industrial activity in rural areas is fast expanding. Furthermore, the relationships between agriculture and industry in rural areas are not always complementary and there are clear tensions and potential conflicts between the two. The challenge, though, is to separate what is good for rural areas (and, more particularly, for rural people), from what is good for agriculture. The prevailing tendency is to see any erosion of agricultural activity and productivity as necessarily anti-developmental in terms of its effects on rural people. But as the discussion in earlier chapters (and above) has indicated, the best interests of rural people may be served if these new activities are permitted – indeed, encouraged – to expand, sometimes *at the expense of agriculture*. The reluctance to entertain this possibility is because of the general failure to examine rural livelihoods in the round.

The notion of blurring also extends into the relationships, and working patterns, of men and women. While Southeast Asia has never had the rigid

---

[6] Jamal and Weeks write this in their reinterpretation of the African situation, but it is just as pertinent to Southeast Asia.

[7] For background to the debate see the following: Lipton (1977), JDS (1984), JDS (1993) and Jamal and Weeks (1993).

gender divisions of labour evident in most other regions of the developing world, these have come under pressure as labour constraints have become more acute. With men and women being selectively enticed into non-farm work, often ex situ, so households have had to contend with quite severe labour shortages at key points in the agricultural cycle. Unable, in many instances, to resort to traditional forms of labour exchange, households have mechanised production, adopted less intensive methods, planted new crops and embraced expedience when it comes to allocating tasks between men and women. Nothing, it seems, is set in stone and traditions are there to be challenged and overturned.

Rural diffferentiation

> ... it is not clear whether agricultural decline and rural poverty result from too much capitalist exploitation or too little (Berry 1993: 137).

Recent work on rural change in Southeast Asia, and in other regions of the developing world, has noted the variable – and changing – bases of rural differentiation. Traditionally, perspectives on wealth, class and social and economic differentiation focused on land. Size of land holdings became, in many studies, a surrogate or proxy for wealth. Large landowners were rich farmers, and small landowners were poor. This, of course, accounts for the traditionally strong link in the rural studies literature between farm (space) and 'farmers' (people) and, in turn, between rural and agricultural. The breaking of this formerly tight association has partially abstracted rural people from rural space. Today, capital (wealth) is often independent of land and human capital has become an increasingly important factor determining avenues and strategies for accumulation. To put it another way, the means of accumulation in the countryside have fundamentally altered for many people. While land and agriculture continue to play a role – and often an important one – this is becoming subservient to other activities.

But this does not mean that we can turn the former association of land with wealth on its head. The new rural rich cannot be equated with those rural people with non-farm income. As the discussion in earlier chapters clearly showed, the fact of non-farm income tells us little about, first, the underlying motivations, second, the means of generation, and third, whether it is broadly developmental from an individual or household perspective. Nonetheless, while reliance on non-farm income may, in some instances, be indicative of extreme poverty, this does not undermine the assertion that access to such income is a critical ingredient allowing rural households to invest in education and income-generating activities (whether farm or non-farm). Rural restructuring, both outwards and upwards (see page 17) has tended to 'decouple the fortunes of farming from the processes of rural development' (Marsden et al. 1990: 2). This is both because the vitality, trajectories of growth and interests of rural areas and rural populations are increasingly independent of farming, and, at the same time, because the interests of farming are increasingly dependent on industries which lie outside agriculture per se (i.e. they lie in the food industry) and often beyond the rural.

Plate 20 Thailand's unemployed elephants (now that a logging ban has put them out of work in the forests) make a living carrying tourists on 'jungle treks'.

Consumption and production

In research on the developed world there has been a good deal of attention paid to the transition of rural spaces from zones of production to places of consumption. Traditionally, rural studies focused almost entirely on issues of production and, in particular, of agricultural production. But as processes of counterurbanisation have become more pronounced, so attention turned to examining how new classes in the countryside, with no links to farming, were altering the balance of power. Consumers of the rural include not only those people who have relocated to the countryside to take advantage of the better 'quality of life' that such living provides, but also those people who live in urban areas and use the countryside for recreation on an occasional basis.

This is not to say that farming has been entirely marginalised. Indeed, it can be argued that even in the developed world the focus in some scholarly quarters on issues of consumption has directed attention away from the fact that farming remains a highly important industry. In the developing world this is, clearly, even more so the case.

But even from the perspective of production, and in areas where agricultural production continues to dominate the rural economy, there are important changes in the farming context which should be acknowledged. Farming needs to be viewed not as a vestigal legacy of some soon-to-be-forgotten world, but a dynamic industry (note: not a 'way of life') which is undergoing continual adaptation. These changes can be viewed as progressing through three stages: first – and most obviously – from subsistence production to commerical production, and subsequently, from commercial production for the food market to commercial production for the food industry. This, as Munton (1992: 28) makes clear, has profoundly changed the place of farming. In progressing to this third stage, farming becomes an integral part of the food industry. Once again, I would not want to push this too far with respect to Southeast Asia. But there is no doubt, whether the focus is on fruit farming for canning factories, contract farming of strawberries and mangetout, or the raising of chickens and pigs for large agro-conglomerates, the farmers in the region are becoming more and more tightly tied into food manufacturing.

Rural change and the global economy

The discussion in this book has emphasised the degree to which locality remains important in determining patterns and directions of change. With the integration of the region, and of rural areas, into wider economic structures it would be misguided to ignore the role of global processes. But when it comes to understanding – beyond the mundane and self-evident – the specifics of what is happening in any given area, then it is the role of local political economies in mediating the global which is critical in bringing precision to any interpretation. This is apparent in the great majority of the case studies highlighted in earlier chapters. It is also clear in Marton's (1998) study of Kunshan in China's Lower Yangzi delta (see page 124).

Marton writes of the 'remarkable spatial transformation in . . . the Chinese countryside . . . [to] open textured landscapes of mixed agricultural and non-agricultural activities' (1998: 3). He argues that the patterns and underlying processes and mechanism of development in the area, while partly determined by exogenous influences are ultimately dictated by 'intensely localized exigencies and opportunities' (1998: 3). Thus, 'place specific economic culture', as he puts it (which embraces factors from local culture to local history), is fundamentally important in setting the framework for change. In Kunshan, local government provided much of the direction and the impetus to the changes which Marton describes. Particularly influential was the Kunshan Rural Industry Bureau. 'Taken together', Marton suggests, 'the administrative, regulatory, and economic functions of the town's industrial corporations were said to provide the "internal engine of development"' (1998: 25). The tight interweaving of agricultural, industrial and residential activities across Kunshan was not viewed as an entirely unmixed blessing. Some local leaders attempted to contain the loss of productive farmland by establishing agricultural protection zones (Marton 1998: 28).[8] Moreover, the importance of local power led to a degree of parochialism – a 'Balkanisation of localities' (1998: 40–41). The potential mismatch between local priorities and regional or national ones is a source of considerable concern. But there is little doubt where the motor of development, and therefore of prosperity lies: in the new rural industries which have colonised the area and rapidly expanded.

The significance of the countryside

A shortcoming of traditional rural studies in geography was a tendency towards a very narrow agrarian focus spanning both economy and society. In the 1970s and 1980s, with industrial political economy at centre stage, rural studies became viewed as a slowly fading outlier in geography. Moreover, rural areas were seen as residual and marginal. This book has suggested that rural areas should neither be viewed as bounded and self-contained, nor as unimportant in terms of the contribution that they can make to wider processes of economic change. Rural areas are, in some instances, at the very centre of economic change and, at the very least, contribute importantly to economic change. Rural areas still exist, but in new ways that are further deconstructing the boundaries between rural and urban, and agriculture and industry.

Much of the debate on rural areas in the developing world has focused on the decline of agriculture and the neglect of the countryside in policy-making. Continuing structural change has pushed agriculture to the margins while governments focus on urban and industrial growth. But there is a case that rural

---

[8] Like scholars working on Southeast Asia, Marton talks of the 'tensions' and 'conflicts' that arose from the pattern of development in Kunshan (1998: 28).

Plate 21   The Regent Hotel in Northern Thailand. The rice fields in the background are contrived for the edification of its guests, giving an illusion of a hotel projected onto a 'timeless' rural space where farmers wear traditional cotton garb and buffaloes prepare the land.

areas are on the verge of finding a new vitality and importance. This is not – as many alternative developmentalists would like – because of a rediscovered respect for traditional farming, but because new classes are taking a direct interest in the countryside, new activities are infiltrating the countryside, and farming itself is restructuring to take account of new opportunities and constraints.

The trajectory of rural change outlined in this book is not irreversible. Agriculture could be revitalised. Intensification of production could be entertained. Non-farm work might lose its current gloss. There is even the possibility that agriculture might, once more, become a desirable occupation and young people might return from their sojourns away from home. This is because farming is a social process embedded in local and national political economies. As such, its progress is contingent on developments in a wide range of arenas from local to global. But while there is the certainty that rural change in Southeast Asia will confound most academics' attempts at prediction, the big picture must be that agriculture and rural life have profoundly changed, and that there is no going back.

Bibliography

ActionAid Vietnam (1995) 'Nine profile villages: Mai Son district, Son La province', unpublished ActionAid report, London.

Akin Rabibhadana (1993) *Social inequity: a source of conflict in the future?*, The 1993 Year End Conference on Who Gets What and How? – Challenges for the Future, Bangkok: Thai Development Research Institute.

Akram-Lodhi, A. Haroon (1998) 'The agrarian question: past and present', *Journal of Peasant Studies* 25(4): 134–49.

Alexander, Jennifer and Alexander, Paul (1982) 'Shared poverty as ideology: agrarian relationships in colonial Java', *Man* 17(4): 597–619.

Alexander, Paul, Boomgaard, Peter and White, Ben (1991) 'Introduction', in: Paul Alexander, Peter Boomgaard, and Ben White (eds) *In the shadow of agriculture: non-farm activities in the Javanese economy, past and present*, Amsterdam: Royal Tropical Institute, pp. 1–13.

Amin, Ash and Thrift, Nigel (1994) 'Living in the global', in: Ash Amin and Nigel Thrift (eds) *Globalization, institutions and regional development in Europe*, Oxford: Oxford University Press, pp. 1–22.

Amsden, Alice H. (1979) 'Taiwan's economic history: a case of étatisme and a challenge to dependency theory', *Modern China* 5(3): 341–80.

Angeles-Reyes, Edna (1994) 'Nonfarm work in the Philippine rural economy: an omen of change or a change of omens?', in: Bruce Koppel, John Hawkins and William James (eds) *Development or deterioration: work in rural Asia*, Boulder, Co and London: Lynne Rienner, pp. 133–65.

Anh Dang, Goldstein, Sidney and McNally, James (1997) 'Internal migration and development in Vietnam', *International Migration Review* 31(2): 312–37.

Antlöv, Hans (1995) *Exemplary centre, administrative periphery: rural leadership and the New Order in Java*, Richmond, Surrey: Curzon Press.

Antlöv, Hans and Svensson, Thommy (1991) 'From rural home weavers to factory labour: the industrialization of textile manufacturing in Majalaya', in: Paul Alexander, Peter Boomgaard and Ben White (eds) *In the shadow of agriculture: nonfarm activities in the Javanese economy, past and present*, Amsterdam: Royal Tropical Institute, pp. 113–26.

Aphichat Chamratrithirong, Kritaya Archavanitkul, Richter, K., Guest, P., Varachai Thongthai, Wathinee Boonchalaksi, Nittaya Piriyathamwoang and Panee Vong-Ek (1995) *National migration survey of Thailand*, Institute for Population and Social Research, Mahidol University, Bangkok.

Apichai Puntasen (1996) 'Agro-industry and self-reliance strategies in village Thailand', in: Mason C. Hoadley, and Christer Gunnarsson (eds) *The village concept in the transformation of rural Southeast Asia, studies from Indonesia, Malaysia and Thailand*, Richmond, Surrey: Curzon Press, pp. 74–91.

Arghiros, Daniel (1997a) 'Rural industry and development in Central Thailand: an alternative approach', *The Journal of Entrepreneurship* 6(1): 1–18.

Arghiros, Daniel (1997b) 'The rise of indigenous capitalists in rural Thailand: profile of brickmakers in the Central Plains', in: Mario Rutten, and Carol Upadhya (eds)

Small business entrepreneurs in Asia and Europe: towards a comparative perspective, New Delhi: Sage, pp. 115–45.

Arghiros, Daniel (1998) 'Mutual obligation and work: changing labour relations in provincial Thai industrial enterprises', paper presented at the EUROSEAS Conference, Hamburg, 3–6 September.

Arghiros, Daniel and Wathana Wongsekiarttirat (1996) 'Development in Thailand's extended metropolitan region: the socio-economic and political implications of rapid change in an Ayutthaya district, central Thailand', in: Michael J.G. Parnwell (ed.) *Thailand: uneven development*, Aldershot: Avebury, pp. 125–45.

Ariffin, Jamilah (1992) *Women and development in Malaysia*, Petaling Jaya, Malaysia: Pelanduk Publications.

Bailey, Warren B. (2000) 'Beware the economic caveman', *Far Eastern Economic Review*, 3 February, p. 29.

Baker, Chris (1999) 'Assembly of the poor: the new drama of village, city and state', paper presented at the 7th International Conference on Thai Studies, University of Amsterdam, 4–8 July.

Banaji, Jarius (1976) 'Summary of selected parts of Kautsky's *The agrarian question*', *Economy and Society* 5(1): 2–49.

Bangkok Post (1998) 'Back to relying on basics', *Bangkok Post Year End Review*, 15 January, pp. 8–9.

Banzon-Bautista, Cynthia (1989) 'The Saudi connection: agrarian change in a Pempangan village, 1977–1984', in: Gillian Hart, Andrew Turton and Benjamin White (eds) *Agrarian transformations: local processes and the state in Southeast Asia*, Berkeley, Ca: University of California Press, pp. 144–58.

Barkin, David (1990) *Distorted development: Mexico in the world economy*, Boulder, Co: Westview Press.

Barnum, Howard N. and Squire, Lyn (1979) 'An econometric application of the theory of the farm-household', *Journal of Development Economics* 6(1): 79–102.

Baxter, Vern and Mann, Susan (1992) 'The survival and revival of non-wage labour in a global economy', *Sociologia Ruralis* 32(2/3): 231–47.

Becker, Gary S. (1991) *A treatise on the family*, Cambridge, Ma: Harvard University Press (enlarged edition).

Bello, Walden (1998) 'The Asian financial crisis: causes, dynamics, and prospects', paper presented at the ASEASUK annual conference, April 1998, London.

Bencha Yoddumnern-Attig (1992) 'Thai family structure and organization: changing roles and duties in historical perspective', in: Bencha Yoddumnern-Attig *et al.* (eds) *Changing roles and statuses of women in Thailand: a documentary assessment*, Institute for Population and Social Research, Mahidol University, Thailand, pp. 8–24.

Bencha Yoddumnern-Attig *et al.* (1992) (eds) *Changing roles and statuses of women in Thailand: a documentary assessment*, Institute for Population and Social Research, Mahidol University, Thailand, pp. 8–24.

Bernstein, Henry (1996) 'Agrarian questions then and now', *Journal of Peasant Studies* 24(1/2): 22–59.

Berry, Sara (1993) *No condition is permanent: the social dynamics of agrarian change in sub-Saharan Africa*, Madison, Wis: University of Wisconsin Press.

Blackwood, Evelyn (1995) 'Senior women, model mothers, and dutiful wives: managing gender contradictions in a Minangkabau village', in: Aihwa Ong and Michael G. Peletz (eds) *Bewitching women, pious men: gender and body politics in Southeast Asia*, Berkeley, Ca: University of California Press, pp. 124–58.

Boomgaard, Peter (1989) *Children of the colonial state: population growth and economic development in Java, 1795–1880*, CASA Monographs no. 1, Amsterdam: Free University Press.

Boomgaard, Peter (1991a) 'The non-agricultural side of an agricultural economy, Java 1500–1900', in: Paul Alexander, Peter Boomgaard and Ben White (eds) *In the shadow of agriculture: non-farm activities in the Javanese economy, past and present*, Amsterdam: Royal Tropical Institute, pp. 14–40.

Boomgaard, Peter (1991b) 'The Javanese village as a Cheshire cat: the Java debate against a European and Latin American background', *Journal of Peasant Studies* 18(2): 288–304.

Booth, Anne (1995) 'Regional disparities and inter-governmental fiscal relations in Indonesia', in: Ian G. Cook, Marcus A. Doel and Rex Li (eds) *Fragmented Asia: regional integration and national disintegration in Pacific Asia*, Aldershot: Avebury, pp. 102–36.

Bowie, Katherine A. (1992) 'Unraveling the myth of the subsistence economy: textile production in nineteenth century Northern Thailand', *Journal of Asian Studies* 51(4): 797–823.

Bray, Francesca (1986) *The rice economies: technology and development in Asian societies*, Oxford: Basil Blackwell.

Breman, Jan (1982) 'The village on Java and the early colonial state', *Journal of Peasant Studies* 9(4): 189–240.

Brow, James (1999) 'Utopia's new-found space: images of the village community in the early writings of Ananda Coomaraswamy', *Modern Asian Studies* 33(1): 67–86.

Brown, Ian (1997) *Economic change in South-East Asia, c.1830–1980*, Kuala Lumpur: Oxford University Press.

Bruner, Edward M. (1961) 'Urbanization and ethnic identity in North Sumatra', *American Anthropologist* 63: 508–521.

Bryceson, Deborah Fahy (1996) 'De-agrarianization and rural employment in sub-Saharan Africa: a sectoral perspective', *World Development* 24(1): 97–111.

Bryceson, Deborah Fahy (1997a) 'De-agrarianisation in sub-Saharan Africa: acknowledging the inevitable', in: Deborah Fahy Bryceson and Vali Jamal (eds) *Farewell to farms: de-agrarianisation and employment in Africa*, Research series 1997/10, African Studies Centre, Leiden, Aldershot: Ashgate.

Bryceson, Deborah Fahy (1997b) 'De-agrarianisation: blessing or blight?', in: Deborah Fahy Bryceson and Vali Jamal (eds) *Farewell to farms: de-agrarianisation and employment in Africa*, Research series 1997/10, African Studies Centre, Leiden, Aldershot: Ashgate, pp. 237–56.

Bryceson, Deborah Fahy and Jamal, Vali (1997) (eds) *Farewell to farms: de-agrarianisation and employment in Africa*, African Studies Centre, Leiden, Aldershot: Ashgate.

Bull, Anna Cento (1987) 'The Lombard silk spinners in the nineteenth century: an industrial workforce in a rural setting', *The Italianist* 7: 99–121.

Bull, Anna Cento (1989) 'Proto-industrialization, small-scale capital accumulation and diffused entrepreneurship: the case of the Brianza in Lombardy (1860–1950)', *Social History* 14(2): 177–200.

Bull, Anna Cento and Corner, Paul (1993) *From peasant to entrepreneur: the survival of the family economy in Italy*, Oxford: Berg.

Bullard, Nicola, Bello, Walden and Malhotra, Kamal (1998) 'Taming the tigers: the IMF and the Asian crisis', *Third World Quarterly* 19(3): 505–55.

Byres, T.J. (1991) 'The agrarian question and differing forms of capitalist agrarian transition: an essay with reference to Asia', in: Jan Breman and Sudipto Mundle (eds) *Rural transformation in Asia*, Delhi: Oxford University Press, pp. 3–76.

Byres, T.J. (1995) 'Political economy, the agrarian question, and the comparative method', *Journal of Peasant Studies* 22(4): 561–80.

Byres, T.J. (1996) *Capitalism from above and capitalism from below: an essay in comparative political economy*, Basingstoke: Macmillan.

Carey, Peter (1986) 'Waiting for the "Just King": the agrarian world of south-central Java from Giyanti (1755) to the Java War (1825–1830)', *Modern Asian Studies* 20(1): 59–137.

Cederroth, Sven (1995) *Survival and profit in rural Java: the case of an East Javanese village*, Richmond, Surrey: Curzon Press.

Chandrasekhar, C.P. (1993) 'Agrarian change and occupational diversification: non-agricultural employment and rural development in West Bengal', *The Journal of Peasant Studies* 20(2): 205–70.

Chang Noi (1997) 'The countryside will save the day', *Thai Development Newsletter* 33 (July–December): 43–5.

Chant, Sylvia (1998) 'Household, gender and rural–urban migration: reflections on linkages and considerations for policy', *Environment and Urbanization* 10(1): 5–21.

Chant, Sylvia and McIlwaine, Cathy (1995a) *Women of a lesser cost: female labour, foreign exchange and Philippine development*, London: Pluto Press.

Chant, Sylvia and McIlwaine, Cathy (1995b) 'Gender and export manufacturing in the Philippines: continuity or change in female employment? The case of the Mactan Export Processing Zone', *Gender, Place and Culture* 2(2): 147–76.

Chantana Banpasirichote (1993) *Community integration into regional industrial development: a case study of Klong Ban Pho Chachoengsao*, Bangkok: Thai Development Research Institute (TDRI).

Chapman, Murray (1995) 'Island autobiographies of movement: alternative ways of knowing?', in: P. Claval, and Singaravelou (eds) *Ethnogéographies*, Paris: L'Harmattan, pp. 247–59.

Chatthip Nartsupha (1986) 'The village economy in pre-capitalist Thailand', in: Seri Phongphit (ed.) *Back to the roots: village and self-reliance in a Thai context*, Bangkok: Rural Development Documentation Centre: pp. 155–65.

Chatthip Nartsupha (1996) 'The village economy in pre-capitalist Thailand', in: Mason C. Hoadley and Christer Gunnarsson (eds) *The village concept in the transformation of rural Southeast Asia, studies from Indonesia, Malaysia and Thailand*, Richmond, Surrey: Curzon Press, pp. 67–73.

Christie, Jan Wisseman (1994) '*Wanua, thani, paraduwan*: the "disintegrating" village in early Java', in: Wolfgang Marschall (ed.) *Texts from the islands: oral and written traditions of Indonesia and the Malay world*, Berne: University of Berne, pp. 27–42.

Chusak Wittayapak (1999) 'The community culture revisited: community as a political space for struggles over natural resources and cultural meaning', paper presented at the 7th International Conference on Thai Studies, University of Amsterdam, 4–8 July.

Cloke, Paul (1996) 'Critical writings on rural studies: a short reply to Simon Miller', *Sociologia Ruralis* 36(1): 117–20.

Cloke, Paul (1998) 'Rural life-styles: material opportunity, cultural experience, and how theory can undermine policy', *Economic Geography* 72: 433–49.

Cloke, Paul and Goodwin, Mark (1992) 'Conceptualizing countryside change: from post-Fordism to rural structured coherence', *Transactions of the Institute of British Geographers (NS)* 17(3): 321–36.

Cohen, Margot (1996) 'Twisting arms for alms', *Far Eastern Economic Review*, 2 May, pp. 25–9.

Courtenay, P.P. (1988) 'Farm size, out-migration and abandoned padi land in Mukim Melekek, Melaka (Peninsular Malaysia)', *Malaysian Journal of Tropical Geography* 17: 18–28.

Cowan, M.P. and Shenton, R.W. (1996) *Doctrines of development*, London: Routledge.

Curry, George and Koczberski, Gina (1998) 'Migration and circulation as a way of life for the Wosera Abelam of Papua New Guinea', *Asia Pacific Viewpoint* 39(1): 29–52.

Curtain, R.L. (1981) 'Migration in Papua New Guinea: the role of the peasant household in a strategy of survival', in: G.W. Jones and H.V. Richter (eds) *Population mobility and development: Southeast Asia and the Pacific*, Australian National University, Canberra: Development Studies Centre, Development Studies Monograph No. 27, pp. 187–204.

Dang Phong (1995) 'Aspects of agricultural economy and rural life in 1993', in: Benedict J. Tria Kerkvliet and Doug J. Porter (eds) *Vietnam's rural transformation*, Boulder, Co: Westview and Singapore: Institute of Southeast Asian Studies, pp. 165–84.

Dasgupta, Biplab (1978) 'Introduction', in: Biplab Dasgupta (ed.) *Village studies in the Third World*, Delhi: Hindustan Publishing Corporation: pp. 1–12.

Davies, Rick and Smith, William (1998) 'The basic necessities survey: the experience of ActionAid Vietnam', London: ActionAid.

Day, Tony (1986) 'How modern was modernity, how traditional tradition, in nineteenth century Java?', *Review of Indonesian and Malaysian Affairs (RIMA)* 20(1): 1–37.

Dearden, Philip (1995) 'Development, the environment and social differentiation in Northern Thailand', in: Jonathan Rigg (ed.) *Counting the costs: economic growth and environmental change in Thailand*, Singapore: ISEAS, pp. 111–30.

De Koninck, Rodolphe (1992) *Malay peasants coping with the world: breaking the community circle?*, Singapore: Institute of Southeast Asian Studies.

Dick, Howard (1993) 'East Java in a regional perspective', in: Howard Dick, James J. Fox and Jamie Mackie (eds) *Balanced development: East Java in the New Order*, Singapore: Oxford University Press, pp. 1–22.

Dick, Howard and Forbes, Dean (1992) 'Transport and communications: a quiet revolution', in: Anne Booth (ed.) *The oil boom and after: Indonesian economic policy and performance in the Soeharto era*, Singapore: Oxford University Press, pp. 258–82.

Dicken, Peter (1994) 'The Roepke lecture in economic geography: global local tensions – firms and states in the global space-economy', *Economic Geography* 70: 101–28.

DiGregorio, Michael R. (1994) *Urban harvest: recycling as a peasant industry in northern Vietnam*, East–West Center Occasional Paper (Environment series) No. 17, Honolulu: East–West Center.

Dixon, Chris (1999) *The Thai economy: uneven development and internationalisation*, London: Routledge.

Douglass, Mike (1998) 'A regional network strategy for reciprocal rural–urban linkages: an agenda for policy research with reference to Indonesia', *Third World Planning Review* 20(1): 1–33.

Drakakis-Smith, David and Dixon, Chris (1997) 'Sustainable urbanization in Vietnam', *Geoforum* 28(1): 21–38.

Economist, The (1999) 'Toiling from there to here', *The Economist*, 31 December 1999, p. 22.

Eder, James F. (1993) 'Family farming and household enterprise in a Philippine community, 1971–1988: persistence or proletarianization?' *Journal of Asian Studies* 52(3): 647–71.

Edmundson, Wade C. (1994) 'Do the rich get richer, do the poor get poorer? East Java, two decades, three villages, 46 people', *Bulletin of Indonesian Economic Studies* 30(2): 133–48.

Effendi, Tadjuddin Noer and Manning, Chris (1994) 'Rural development and nonfarm employment in Java', in: Bruce Koppel, John Hawkins and William James (eds) *Development or deterioration: work in rural Asia*, Boulder, Co and London: Lynne Rienner, pp. 211–47.

Ellis, Frank (1998) 'Household strategies and rural livelihood diversification', *Journal of Development Studies* 35(1): 1–38.

Elmhirst, Rebecca (1995) 'Gender, environment and transmigration: comparing migrant and *pribumi* household strategies in Lampung, Indonesia', paper presented to the Third WIVS conference on Indonesian Women in the Household and Beyond, Royal Institute of Linguistics and Anthropology, Leiden, 25–29 September.

Elmhirst, Rebecca (1996) 'Transmigration and local communities in North Lampung: exploring identity politics and resource control in Indonesia', paper presented at the Association of South East Asian Studies' (ASEASUK) Conference, School of Oriental & African Studies, London, 25–27 April.

Elmhirst, Rebecca (1997) 'Gender, environment and culture: a political ecology of transmigration in Indonesia', PhD thesis, Wye College, University of London, London.

Elmhirst, Rebecca (1998a) 'Daughters and displacement: migration dynamics in an Indonesian transmigration area', paper presented at the workshop on Migration and Sustainable Livelihoods, University of Sussex, 5–6 June.

Elmhirst, Rebecca (1998b) 'Gender, culture and space: a political geography of factory labour in Indonesia', paper presented at the European South East Asian Studies (EUROSEAS) conference, Hamburg, 3–6 September.

Elson, Robert E. (1986) 'Aspects of peasant life in early 19th century Java', in: D. Chandler and M.C. Ricklefs (eds) *Nineteenth and twentieth century Indonesia*, Clayton, Victoria: Centre for Southeast Asian Studies, pp. 57–81.

Elson, Robert E. (1992) 'International commerce, the state and society: economic and social change', in: Nicholas Tarling (ed.) *The Cambridge history of Southeast Asia: the nineteenth and twentieth centuries* (volume II), Cambridge: Cambridge University Press.

Elson, Robert E. (1997) *The end of the peasantry in Southeast Asia: a social and economic history of peasant livelihood, 1800–1990s*, Basingstoke: Macmillan.

Escobar, Arturo (1995) *Encountering development: the making and unmaking of the Third World*, Princeton, NJ: Princeton University Press.

Estudillo, Jonna P. and Otsuka, Keijiro (1999) 'Green revolution, human capital and off-farm employment: changing sources of income among farm households in Central Luzon, 1966–1994, *Economic Development and Cultural Change* 47(3): 497–523.

Evans, Hugh Emrys (1992) 'A virtuous circle model of rural–urban development: evidence from a Kenyan small town and its hinterland', *Journal of Development Studies* 28(4): 640–67.

Evans, Hugh Emrys and Ngau, Peter (1991) 'Rural–urban relations, household income diversification, and agricultural productivity', *Development and Change* 22: 519–45.

Fairclough, Gordon (1995) 'Expensive and difficult: voting rules are tough on migrant workers', *Far Eastern Economic Review*, 6 July, p. 17.

Falkus, Malcolm and Hewison, Kevin (1999) 'Thailand's crisis: economic explanations and political responses', paper presented to the 7th International Conference on Thai Studies, Amsterdam, 4–8 July.

Fall, Abdou Salam (1998) 'Migrants' long-distance relationships and social networks in Dakar', *Environment and Urbanization* 10(1): 135–45.

Fegan, Brian (1983) 'Establishment fund, population increase and changing class structures in Central Luzon', *Philippine Sociological Review* 31: 31–43.

Firman, Tommy (1991) 'Rural households, labor flows and the housing construction industry in Bandung, Indonesia', *Tijdschrift voor Economische en Sociale Geografie* 82(2): 94–105.

Firman, Tommy (1994) 'Labour allocation, mobility, and remittances in rural households: a case from Central Java, Indonesia', *Sojourn* 9(1): 81–101.

Fisher, Thomas and Mahajan, Vijay with Singha, Ashok (1997) *The forgotten sector: non-farm employment and rural enterprises in rural India*, London: Intermediate Technology Publications.

Flaherty, Mark, Vandergeest, Peter and Miller, Paul (1999) 'Rice paddy or shrimp pond: tough decisions in rural Thailand', *World Development* 27(12): 2045–60.

Floro, Maria Sagrario and Schaefer, Kendall (1998) 'Restructuring of the labor markets in the Philippines and Zambia: the gender dimension', *Journal of Developing Areas* 33(1): 73–98.

Folbre, Nancy (1984) 'Household production in the Philippines: a non-neoclassical approach', *Economic Development and Cultural Change* 32(2): 303–30.

Folbre, Nancy (1986) 'Cleaning house: new perspectives on households and economic development', *Journal of Development Economics* 22(1): 5–40.

Fox, James (1977) *Harvest of the palm: ecological change in eastern Indonesia*, Cambridge, Ma: Harvard University Press.

Fox, James (1991) 'Managing the ecology of rice production in Indonesia', in: Joan Hardjono (ed.) *Indonesia: resources, ecology and environment*, Singapore: Oxford University Press, pp. 61–84.

Fox, James (1993) 'The rice baskets of East Java: the ecology and social context of *sawah* production', in: Howard Dick, James J. Fox and Jamie Mackie (eds) *Balanced development: East Java in the New Order*, Singapore: Oxford University Press, pp. 120–57.

Francis, Elizabeth (1998) 'Gender and rural livelihoods in Kenya', *Journal of Development Studies* 35(2): 72–95.

Francis, Elizabeth and Hoddinott, John (1993) 'Migration and differentiation in western Kenya: a tale of two sub-locations', *Journal of Development Studies* 30(1): 115–45.

Francks, Penelope with Boestal, Johanna and Choo Hyop Kim (1999) *Agriculture and economic development in East Asia: from growth to protectionism in Japan, Korea and Taiwan*, London: Routledge.

Friedmann, Harriet (1993) 'The political economy of food: a global crisis', *New Left Review* 197 (Jan/Feb): 29–57.

Funahashi, Kazuo (1996) 'Farming by the older generation: the exodus of young labor in Yasothon province, Thailand', *Tonan Ajia Kenky (Southeast Asian Studies)* 33(4): 107–21.

Geertz, Clifford (1963) *Agricultural involution: the process of ecological change in Indonesia*, Berkeley, Ca: University of California Press.

Ginsburg, Norton (1991) 'Extended metropolitan regions in Asia: a new spatial paradigm', in: Norton Ginsburg, Bruce Koppel and T.G. McGee (eds) *The extended metropolis: settlement transition in Asia*, Honolulu: University of Hawaii Press, pp. 27–46.

Goodman, David (1997) 'World-scale processes and agro-food systems: critique and research needs', *Review of International Political Economy* 4(4): 663–87.

Goodman, David, Sorj, Bernardo and Wilkinson, John (1987) *From farming to biotechnology: a theory of agro-industrial development*, Oxford: Basil Blackwell.

Goodman, David and Watts, Michael (1994) 'Reconfiguring the rural or Fording the divide?: Capitalist restructuring and the global agro-food system', *Journal of Peasant Studies* 22(1): 1–49.

Grabowski, Richard (1995) 'Commercialization, nonagricultural production, agricultural innovation, and economic development', *Journal of Developing Areas* 30: 41–62

Grandstaff, Terry (1988) 'Environment and economic diversity in Northeast Thailand', in: Terd Charoenwatana and A. Terry Rambo (eds) *Sustainable rural development in Asia*, Khon Kaen, Thailand: Khon Kaen University, pp. 11–22.

Grandstaff, Terry (1992) 'The human environment: variation and uncertainty', *Pacific Viewpoint* 33(2): 135–44.

Guinness, Patrick (1994) 'Local society and culture' in: Hal Hill (ed.) *Indonesia's New Order: the dynamics of socio-economic transformation*, Honolulu: University of Hawaii Press, pp. 267–304.

Guinness, Patrick and Husin, Imron (1993) 'Industrial expansion into a rural subdistrict: Pandaan, East Java', in: Howard Dick, James J. Fox and Jamie Mackie (eds) *Balanced development: East Java in the New Order*, Singapore: Oxford University Press, pp. 272–95.

Guldin, Gregory Eliya (1994) 'Townization and Civilization in Southern China: desakotas and beyond'. Paper presented to the American Anthropological Association, Atlantic, GA.

Haggblade, Steven, Hazell, Peter and Brown, James (1989) 'Farm–nonfarm linkages in rural sub-Saharan Africa', *World Development* 17(8): 1173–201.

Halfacree, Keith H. (1994) 'The importance of "the rural" in the constitution of counter urbanization: evidence from England in the 1980s', *Sociologia Ruralis*, 34(2–3): 164–89.

Hart, Gillian (1986) *Power, labor and livelihood: processes of change in rural Java*, Berkeley, Ca: University of California Press.

Hart, Gillian (1992) 'Household production reconsidered: gender, labor conflict, and technological change in Malaysia's Muda region', *World Development* 20(6): 809–23.

Hart, Gillian (1994) 'The dynamics of diversification in an Asian rice region', in: Bruce Koppel, John Hawkins and William James (eds) *Development or deterioration: work in rural Asia*, Boulder Co and London: Lynne Rienner, pp. 47–71.

Hart, Gillian (1995) 'Gender and household dynamics: recent theories and their implications', in: M.G. Quibria (ed.) *Critical issues in Asian development: theories, experiences and policies*, Hong Kong: Oxford University Press with the Asian Development Bank, pp. 39–74.

Hart, Gillian (1996) 'The agrarian question and industrial dispersal in South Africa: agro-industrial linkages through Asian lenses', *Journal of Peasant Studies* 23(2–3): 245–77.

Hart, Gillian (1997) 'Multiple trajectories of rural industrialisation: an agrarian critique of industrial restructuring and the new institutionalism', in: David Goodman and Michael Watts (eds) *Globalising food: agrarian questions and global restructuring*, London: Routledge, pp. 56–78.

Hart, Gillian, Turton, Andrew and White, Benjamin (1989) 'Introduction' in Gillian Hart, Andrew Turton and Benjamin White (eds) *Agrarian transformations: local processes and the state in Southeast Asia*, Berkeley, Ca: University of California Press, pp. 1–11.

Hayami, Yujiro and Kikuchi, Masao (1981) *Asian village economy at the crossroads: an economic approach to institutional change*, Tokyo: University of Tokyo Press.

Hayami, Yujiro, Kikuchi, Masao and Marciano, Esther B. (1998) 'Structure of rural-based industrialization: metal craft manufacturing on the outskirts of greater Manila, the Philippines', *The Developing Economies* 36(2): 132–54.

Healey, Chris (1996) 'Aru connections: outback Indonesia in the modern world', in: David Mearns and Chris Healey (eds) *Remaking Maluku: social transformation in Eastern*

Indonesia, Special monograph no. 1, North Territory University, Darwin: Centre for Southeast Asian Studies, pp. 14–26.

Hefner, Robert W. (1990) *The political economy of mountain Java: an interpretive history*, Berkeley, Ca: University of California Press.

Herdt, Robert W. (1987) 'A retrospective view of technological and other change in Philippine rice farming, 1965–1982', *Economic Development and Cultural Change* 35: 329–49.

Hetler, Carol B. (1989) 'The impact of circular migration on a village economy', *Bulletin of Indonesian Economic Studies* 25(1): 53–75.

Hettne, Bjorn (1990) *Development theory and the three worlds*, Harlow: Longman.

Hettne, Bjorn (1995) *Development theory and the three worlds: towards an international political economy of development* (2nd edition), Harlow: Longman.

Hewison, Kevin (1999) 'Localism in Thailand: a study of globalisation and its discontents', CSGR working paper no. 39/99, University of Warwick: Centre for the Study of Globalisation and Regionalisation (http:///www.csgr.org).

Hiebert, Murray (1993) 'A fortune in waste: scarcity forces Vietnam to reuse its resources', *Far Eastern Economic Review*, 23 December, p. 36.

Hill, Hal (1996) *The Indonesian economy since 1996: Southeast Asia's emerging giant*, Cambridge: Cambridge University Press.

Hirsch, Philip (1989a) 'Local contexts of differentiation and inequality on the Thai periphery', *Journal of Contemporary Asia* 19(3): 308–23.

Hirsch, Philip (1989b) 'The state in the village: interpreting rural development in Thailand', *Development and Change* 20(1): 35–56.

Hirsch, Philip (1990a) *Development dilemmas in rural Thailand*, Singapore: Oxford University Press.

Hirsch, Philip (1990b) 'Forests, forest reserve, and forest land in Thailand', *The Geographical Journal* 156(2): 166–74.

Hirsch, Philip (1992) 'State, capital, and land in recently cleared areas of western Thailand', *Pacific Viewpoint* 33(1): 36–57.

Hoadley, Mason C. (1996) 'Non-village political economy of pre-colonial West Java', in: Mason C. Hoadley and Christer Gunnarsson (eds) *The village concept in the transformation of rural Southeast Asia, studies from Indonesia, Malaysia and Thailand*, Richmond, Surrey: Curzon Press, pp. 29–43.

Hoadley, Mason C. and Gunnarsson, Christer (1996) 'Introduction', in: Mason C. Hoadley and Christer Gunnarsson (eds) *The village concept in the transformation of rural Southeast Asia, studies from Indonesia, Malaysia and Thailand*, Richmond, Surrey: Curzon Press, pp. vii-xviii.

Hobart, Mark (1993) 'Introduction: the growth of ignorance', in Mark Hobart (ed.) *An anthropological critique of development: the growth of ignorance*, London: Routledge, pp. 1–30.

Hobsbawm, Eric (1994) *Age of extremes: the short twentieth century 1914–1991*, London: Michael Joseph.

Hoggart, Keith (1990) 'Let's do away with the rural', *Journal of Rural Studies* 6(3): 245–57.

Hoggart, Keith (1992) 'Global economic structures and agricultural changes', in: Keith Hoggart (ed.) *Agricultural change, environment and economy: essays in honour of W.B. Morgan*, London: Mansell, pp. 1–24.

Hoggart, Keith (1997) 'Rural migration and counter-urbanization in the European periphery: the case of Andalucía', *Sociologia Ruralis* 37(1): 134–53.

Hugo, Graeme (1993) 'Indonesian labour migration to Malaysia: trends and policy implications', *Southeast Asian Journal of Social Science* 21(1): 36–70.

Hugo, Graeme (1997) 'Asia and the Pacific on the move: workers and refugees, a challenge to nation states', *Asia Pacific Viewpoint* 38(3): 267–86.

Hüsken, Frans (1981) *Regional diversity in Javanese agrarian development: variations in the pattern of involution*, Working paper no. 10, Sociology of Development Research Centre, University of Bielefeld, Bielefeld.

Ikemoto, Yukio (1996) 'Expansion of cottage industry in Northeast Thailand: the case of triangular pillows in Yasothon province', *Tonan Ajia Kenky (Southeast Asian Studies)* 33(4): 122–37.

Ilbery, Brian (1998a) (ed.) *The geography of rural change*, Harlow: Addison Wesley Longman.

Ilbery, Brian (1998b) 'Dimensions of rural change', in Brian Ilbery (ed.) *The geography of rural change*, Harlow: Addison Wesley Longman, pp. 1–10.

Jamal, Vali and Weeks, John (1993) *Africa misunderstood or whatever happened to the rural–urban gap?*, Basingstoke: Macmillan.

Jamieson, Neil (1991) 'The dispersed metropolis in Asia: attitudes and trends in Java', in: Norton Ginsburg, Bruce Koppel and T.G. McGee (eds) *The extended metropolis: settlement transition in Asia*, Honolulu: University of Hawaii Press, pp. 275–97.

JDS (1984) 'Development and the rural–urban divide' (special issue), *Journal of Development Studies* 20(3).

JDS (1993) 'Beyond urban bias' (special issue), *Journal of Development Studies* 29(4).

Jefremovas, Villia (1992) 'Gender, the household and cash cropping in Sagada, the Philippines: initial impressions from the field', in: Penny Van Esterik and John Van Esterik (eds) *Gender and development in Southeast Asia*, Montreal: Canadian Council for Southeast Asian Studies, pp. 51–8.

Jervell, Anne Moxnes (1999) 'Changing patterns of family farming and pluriactivity', *Sociologia Ruralis* 39(1): 100–16.

Jirström, Magnus (1996) *In the wake of the green revolution: environmental and socio-economic consequences of intensive rice agriculture*, Lund, Sweden: Lund University Press.

Johnston, D.C. (1998) 'These roads were made for walking? Provision of rural transport services in Indonesia', paper presented at the 5th International Conference on Southeast Asian Geography, Singapore, 30 November–4 December 1998.

Jones, Huw and Tieng Pardthaisong (1999) 'The impact of overseas labour migration on rural Thailand: regional, community and individual dimensions', *Journal of Rural Studies*, 15(1): 35–47.

Kalantaridis, Christos and Labrianidis, Lois (1999) 'Family production and the global market: rural industrial growth in Greece', *Sociologia Ruralis* 39(2): 146–64.

Kamete, Amin Y. (1998) 'Interlocking livelihoods: farm and small town in Zimbabwe', *Environment and Urbanization* 10(1): 23–34.

Kato, Tsuyoshi (1994) 'The emergence of abandoned paddy fields in Negeri Sembilan, Malaysia', *Tonan Ajia Kenky (Southeast Asian Studies)* 32(2): 145–72.

Kaufman, Howard K. (1977) *Bangkhuad: a community study in Thailand*, Rutland, Vt: Charles Tuttle.

Kearney, Michael (1996) *Reconceptualizing the peasantry: anthropology in global perspective*, Boulder Co: Westview Press.

Kelly, Philip (1998) 'The politics of urban-rural relations: land use conversion in the Philippines', *Environment and Urbanization* 10(1): 35–54.

Kelly, Philip (1999a) 'Everyday urbanization: the social dynamics of development in Manila's extended metropolitan region', *International Journal of Urban and Regional Research* 23(2): 283–303.

Kelly, Philip (1999b) 'Rethinking the "local" in labour markets: the consequences of cultural embeddedness in a Philippine growth zone', *Singapore Journal of Tropical Geography* 20(1): 56–75.

Kemp, Jeremy (1988) *Seductive mirage: the search for the village community in Southeast Asia*, Dordrecht: Foris.

Kemp, Jeremy (1989) 'Peasants and cities: the cultural and social image of the Thai peasant community', *Sojourn* 4(1): 6–19.

Kemp, Jeremy (1991) 'The dialectics of village and state in modern Thailand', *Journal of Southeast Asian Studies* 22(2): 312–26.

King, Victor T. (1999) *Anthropology and development in South-East Asia*, Kuala Lumpur: Oxford University Press.

Kirkby, Richard and Zhao Xiaobin (1999) 'Sectoral and structural considerations in China's rural development', *Tijdschrift voor Economische en Sociale Geografie* 90(3): 272–84.

Kitahara, Atsushi (1996) *The Thai rural community reconsidered: historical community formation and contemporary development movements*, The Political Economy Centre, Faculty of Economics, Chulalongkorn University, Bangkok.

Klopfer, Lisa (1994) *Confronting modernity in a rice-producing community: contemporary values and identity among the highland Minangkabau of West Sumatra, Indonesia*, Ann Arbor, Mi: UMI.

Knight, G.R. (1982) 'Capitalism and commodity production in Java', in Hamza Alavi, P.L. Burns, G.R. Knight, P.B. Mayer and Doug McEachern (eds) *Capitalism and colonial production*, London: Croom Helm, pp. 119–58.

Koizumi, Junko (1992) 'The commutation of Suai from Northeast Siam in the middle of the nineteenth century', *Journal of Southeast Asian Studies* 23(2): 276–307.

Konchan, Somkiat and Kono, Yasuyuki (1996) 'Spread of direct seeded lowland rice in Northeast Thailand: farmers' adaptation to economic growth', *Tonan Ajia Kenky (Southeast Asian Studies)* 33(4): 5–19.

Koopman, Jeanne (1991) 'Neoclassical household models and modes of household production: problems in the analysis of African agricultural households', *Review of Radical Political Economics* 23(3&4): 148–73.

Koppel, Bruce and Hawkins, John (1994) 'Rural transformation and the future of work in rural Asia', in: Bruce Koppel, John Hawkins and William James (eds) *Development or deterioration: work in rural Asia*, Boulder, Co and London: Lynne Rienner, pp. 1–46.

Koppel, Bruce and James, William (1994) 'Development or deterioration? Understanding employment diversification in rural Asia', in: Bruce Koppel, John Hawkins and William James (eds) *Development or deterioration: work in rural Asia*, Boulder, Co and London: Lynne Rienner, pp. 275–301.

Krüger, Fred (1998) 'Taking advantage of rural assets as a coping strategy for the rural poor: the case of rural–urban interrelations in Botswana', *Environment and Urbanization* 10(1): 119–34.

Kunsiri Olarikkachat (1998) 'Thais say: free us from this thirst for affluence', *Bangkok Post*, 2 February 1998, http://www.bkkpost.samart.co/th/news/BParchive/BP19980202/020298_News04.html.

Kunsiri Olarikkachat and Ampa Santimatanedol (1997) 'Royal call for self-reliance wins support', *Bangkok Post*, 30 December 1997, http://www.bkkpost.samart.co/th/news/BParchive/BP971230/301297_News18.html.

Lanjouw, Peter (1999) 'Rural nonagricultural employment and poverty in Ecuador', *Economic Development and Cultural Change* 48(1): 91–122.

Leaf, M. (1996) 'Building the road for the BMW: culture, vision, and the extended metropolitan region of Jakarta', *Environment and Planning A* 28: 1617–1635.

Leinbach, Thomas R. and Del Casino, Vincent J. Jr. (1998) 'The family mode of production and its fungibility in Indonesian transmigration: the example of Makarti Jaya, South Sumatra', *Sojourn* 13(2): 193–219.

Leinbach, Thomas R. and Watkins, John E. (1998) 'Remittances and circulation behaviour in the livelihood process: transmigrant families in South Sumatra, Indonesia', *Economic Geography* 71(1): 45–63.

Leones, J.P. and Feldman, S. (1998) 'Nonfarm activity and rural household income: evidence from Philippines microdata', *Economic Development and Cultural Change* 46: 789–806.

Li, Tania Murray (1996) 'Household formation, private property, and the state', *Sojourn* 11(2): 259–87.

Lipton, Michael (1977) *Why poor people stay poor: urban bias in world development*, London: Temple Smith.

Little, Peter D. and Watts, Michael J. (1994) (eds) *Living under contract: contract farming and agrarian transformation in sub-Saharan Africa*, Madison, Wis: University of Wisconsin Press.

Lok, Helen (1993) 'Labour in the garment industry: an employer's perspective', in: Chris Manning and Joan Hardjono (eds) *Indonesia assessment 1993–Labour: sharing in the benefits of growth?*', Political and Social Change Monograph no. 20, Research School of Pacific Studies, Canberra: Australian National University, pp. 155–72.

Luxmon Wongsuphasawat (1995) 'The extended metropolitan region and uneven industrial development in Thailand', paper presented at the first EUROSEAS Conference, Leiden, 29 June–1 July 1995.

MacPhail, Fiona (1992) 'Household decision-making in South Sulawesi, Indonesia: complex reality versus fancy (economic) models', in: Penny Van Esterik and John Van Esterik (eds) *Gender and development in Southeast Asia*, Montreal: Canadian Council for Southeast Asian Studies, pp. 31–50.

Macpherson, Cluny (1985) 'Public and private views of home: will Western Samoan migrants return?', *Pacific Viewpoint* 26(2): 242–62.

Macpherson, Cluny (1994) 'Changing patterns of commitment to island homelands: a case study of Western Samoa', *Pacific Studies* 17(3): 83–116.

Mantra, Ida Bagoes (1999) 'Illegal Indonesian labour movement from Lombok to Malaysia', *Asia Pacific Viewpoint* 40(1): 59–68.

Marsden, Terry (1998) 'Economic perspectives', in: Brian Ilbery (ed.) *The geography of rural change*, Harlow: Addison Wesley Longman, pp. 13–30.

Marsden, Terry, Lowe, Philip and Whatmore, Sarah (1990a) (eds) *Rural restructuring: global processes and their responses*, London: David Fulton.

Marsden, Terry, Lowe, Philip and Whatmore, Sarah (1990b) 'Introduction: questions of rurality', in: Terry Marsden, Philip Lowe and Sarah Whatmore (eds) *Rural restructuring: global processes and their responses*, London: David Fulton, pp. 1–20.

Marsden, Terry Munton, Richard, Ward, Neil and Sarah Whatmore (1996) 'Agricultural geography and the political economy approach', *Economic Geography* 72: 361–75.

Marsden, Terry, Murdoch, Jonathan, Lowe, Philip, Munton, Richard and Flynn, Andrew (1993) *Constructing the countryside*, London: UCL Press.

Marton, Andrew M. (1998) *Urbanization in China's Lower Yangzi Delta: transactional relations and the repositioning of locality*, East Asian Institute Occasional Paper no. 10, Singapore: Singapore University Press.

Mason, Andrew D. (1996) 'Targeting the poor in rural Java', *IDS Bulletin* 27(1): 67–82.

Mather, Celia E. (1983) 'Industrialization in the Tangerang Regency of West Java: women workers and the Islamic patriarchy', *Bulletin of Concerned Asian Scholars* 15(2): 2–17.

Maurer, Jean-Luc (1991) 'Beyond the sawah: economic diversification in four Bantul villages, 1972–1987', in: Paul Alexander, Peter Boomgaard, and Ben White (eds) *In the shadow of agriculture: non-farm activities in the Javanese economy, past and present*, Amsterdam: Royal Tropical Institute, pp. 92–112.

McGee, T.G. (1989) 'Urbanisasi or kotadesasi? Evolving patterns of urbanization in Asia', in: Frank. J. Costa, Ashok K. Dutt, Lawrence J.C. Ma and Allen G. Noble (eds) *Urbanization in Asia: spatial dimensions and policy issues*, Honolulu: University of Hawaii Press, pp. 93–108.

McGee, T.G. (1991) 'The emergence of desakota regions in Asia: expanding a hypothesis', in: Norton Ginsburg, Bruce Koppel and T.G. McGee (eds) *The extended metropolis: settlement transition in Asia*, Honolulu: University of Hawaii Press, pp. 3–25.

McGee, T.G. and Greenberg, Charles (1992) 'The emergence of extended metropolitan regions in ASEAN', *ASEAN Economic Bulletin* 9 (1): 22–44.

McGee, Terry (1995) 'The urban future of Vietnam', *Third World Planning Review* 17(3): 253–77.

McMichael, Philip (1997) 'Rethinking globalization: the agrarian question revisited', *Review of International Political Economy* 4(4): 630–62.

McMichael, Philip and Myhre, David (1991) 'Global regulation vs. the nation state: agro-food systems and the new politics of capital', *Capital and Class* 43 (spring): 83–105.

McVey, Ruth (1992) 'The materialization of the Southeast Asian entrepreneur', in: Ruth McVey (ed.) *Southeast Asian capitalists*, Studies on Southeast Asia, Ithaca, NY: Cornell University Press, pp. 7–33.

Meagher, Kate and Mustapha, Abdul Raufu (1997) 'Not by farming alone: the role of non-farm incomes in rural Hausaland', in: Deborah Fahy Bryceson and Vali Jamal (eds) *Farewell to farms: de-agrarianisation and employment in Africa*, Research series 1997/10, African Studies Centre, Leiden, Aldershot: Ashgate, pp. 63–84.

Miller, Simon (1996) 'Class, power and social construction: issues of theory and application in thirty years of rural studies', *Sociologia Ruralis* 36(1): 93–116.

Mills, Mary Beth (1997) 'Contesting the margins of modernity: women, migration, and consumption in Thailand', *American Ethnologist* 24(1): 37–61.

Miyagawa, Shuichi (1996) 'Recent expansion of non-glutinous rice cultivation in Northeast Thailand: intraregional variation', *Tonan Ajia Kenky (Southeast Asian Studies)* 33(4): 29–56.

MOAC (1989) *Agricultural statistics of Thailand crop year 1988/89*, Agricultural statistics no. 414, Bangkok: Ministry of Agriculture and Cooperatives.

Moerman, Michael (1968) *Agricultural change and peasant choice in a Thai village*, Berkeley, Ca: University of California Press.

Moerman, Michael and Miller, Patricia (1989) 'Changes in a village's relations with its environment', in: *Culture and environment in Thailand: a symposium of the Siam Society*, Bangkok: Siam Society, pp. 303–26.

Monk, Janice and Katz, Cindi (1993) 'When in the world are women?', in: Cindi Katz and Janice Monk (eds) *Full circles: geographies of women over the life course*, London: Routledge.

Moran, Warren, Blunden, Greg and Greenwood, Julie (1993) 'The role of family farming in agrarian change', *Progress in Human Geography* 17(1): 22–42.

Mormont, Marc (1990) 'Who is rural? or, how to be rural: towards a sociology of the rural', in: Terry Marsden, Philip Lowe and Sarah Whatmore (eds) *Rural restructuring: global processes and their responses*, London: David Fulton, pp. 21–44.

Muijzenberg, Otto van den (1991) 'Tenant emancipation, diversification and social differentiation in Central Luzon', in: Jan Breman and Sudipto Mundle (eds) *Rural transformation in Asia*, Delhi: Oxford University Press, pp. 313–37.

Munton, Richard J.C. (1992) 'The uneven development of capitalist agriculture: the repositioning of agriculture within the food system', in: Keith Hoggart (ed.) *Agricultural change, environment and economy: essays in honour of W.B. Morgan*, London: Mansell, pp. 25–48.

Murray, Alison J. (1991) *No money, no honey: a study of street traders and prostitutes in Jakarta*, Singapore: Oxford University Press.

Naylor, Rosamond (1992) 'Labour-saving technologies in the Javanese rice economy: recent developments and a look into the 1990s', *Bulletin of Indonesian Economic Studies* 28(3): 71–89.

Nipon Poapongsakorn (1994) 'Transformations in the Thai rural labor market', in: Bruce Koppel, John Hawkins and William James (eds) *Development or deterioration: work in rural Asia*, Boulder, Co and London: Lynne Rienner, pp. 167–210.

NSO (1993) *Statistical handbook of Thailand 1993*, Bangkok: National Statistical Office.

Oey-Gardiner, Mayling (1991) 'Gender differences in schooling in Indonesia', *Bulletin of Indonesian Economic Studies* 27(1): 57–79.

Ong, Aihwa (1987) *Spirits of resistance and capitalist discipline: factory women in Malaysia*, New York: SUNY Press.

Otsuka, Keijiro, Gascon, Fe and Asano, Seki (1994) 'Green revolution and labour demand in rice farming: the case of central Luzon, 1966–1990', *Journal of Development Studies* 31(1): 82–109.

Overton, John (1994) *Colonial Green Revolution? Food, irrigation and the state in colonial Malaya*, Wallingford: CAB International.

Owen, Norman G. (1987) (ed.) *Death and disease in Southeast Asia: explorations in social, medical and demographic history*, Singapore: Oxford University Press.

Page, Brian (1996) 'Across the great divide: agriculture and industrial geography', *Economic Geography* 72: 376–97.

Pandey, S. and Velasco, L. (1999) 'Economics of direct seeding in Asia: patterns of adoption and research priorities', *International Rice Research Notes* 24(2): 6–11.

Parish, William L., Xiaoye Zhe and Fang Li (1995) 'Nonfarm work and marketization of the Chinese countryside', *China Quarterly* 143: 697–730.

Paritta Chalermpow Koanantakool and Askew, Marc (1993) *Urban life and urban people in transition*, The 1993 Year End Conference on Who Gets What and How? – Challenges for the Future, Bangkok: TDRI.

Parnwell, Michael (1986) 'Migration and the development of agriculture: a case study of North-East Thailand', in: Michael J.G. Parnwell (ed.) *Rural development in North-East Thailand: case studies of migration, irrigation and rural credit*, Occasional paper no. 12, Centre for South East Asian Studies, University of Hull, pp. 93–140.

Parnwell, Michael J.G. (1990) *Rural industrialisation in Thailand*, Hull Paper in Developing Area Studies no. 1, Centre of Developing Area Studies, University of Hull.

Parnwell, Michael J.G. (1993) 'Tourism, handicrafts and development in North-East Thailand', paper presented at the 5th International Thai Studies Conference, SOAS, London, July.

Parnwell, Michael J.G. (1994) 'Rural industrialisation and sustainable development in Thailand', *Thai Environment Institute Quarterly Environment Journal* 1(2): 24–39.

Parnwell, Michael J.G. and Arghiros, Daniel (1996) 'Uneven development in Thailand', in: Michael Parnwell (ed.) *Uneven development in Thailand*, Aldershot: Avebury, pp. 1–27.

Parnwell, Michael J.G. and Luxmon Wongsuphasawat (1997) 'Between the global and the local: extended metropolitanisation and industrial location decision making in Thailand', *Third World Planning Review* 19(2): 119–38.

Parnwell, Michael J.G. and Suranart Khamanarong (1996) 'Rural industrialisation in Thailand: village industries as a potential basis for rural development in the North-East', in: Michael J.G. Parnwell (ed.) *Thailand: uneven development*, Aldershot: Avebury, pp.161–85.

Pasuk Phongpaichit (1984) 'The Bangkok masseuses: origins, status and prospects', in: G.W. Jones (ed.) *Women in the urban and industrial workforce: Southeast and East Asia*, Canberra: Australian National University.

Pasuk Phongpaichit and Baker, Chris (1998) *Thailand's boom and bust*, Chiang Mai: Silkworm Books.

Paulson, Deborah D. and Rogers, Steve (1997) 'Maintaining subsistence security in Western Samoa', *Geoforum* 28(2): 173–87.

Phillips, Martin (1998) 'Social perspectives', in: Brian Ilbery (ed.) *The geography of rural change*, Harlow: Addison Wesley Longman, pp. 31–54.

Pincus, Jonathan (1996) *Class power and agrarian change: land and labour in rural West Java*, Basingstoke: Macmillan.

Ploeg, Jan Douwe van der (1993) 'Rural sociology and the new agrarian question', *Sociologia Ruralis* 33(2): 240–60.

Popkin, Samuel L. (1979) *The rational peasant: the political economy of rural society in Vietnam*, Berkeley, Ca: University of California Press.

Preston, David A. (1989) 'Too busy to farm: under-utilisation of farm land in Central Java', *Journal of Development Studies* 26(1): 43–57.

Preston, David A. (1994) 'Rapid household appraisal: a method for facilitating the analysis of household livelihood strategies', *Applied Geography* 14: 203–13.

Preston, David A. (1998) 'Changed household livelihood strategies in the Cordillera of Luzon', *Tijdschrift voor Economische en Sociale Geografie* 89(4): 371–83.

Radcliffe, Sarah A. (1986) 'Gender relations, peasant livelihood strategies and migration: a case study from Cuzco, Peru', *Bulletin of Latin American Research* 5(2): 29–47.

Rambo, A. Terry (1977) 'Closed corporate and open peasant communities: reopening a hastily shut case', *Comparative Studies in Society and History* 19(2): 179–88.

Rawski, Thomas G. and Mead, Robert W. (1998) 'On the trail of China's phantom farmers', *World Development* 26(5): 767–81.

Raynolds, Laura (1997) 'Restructuring national agriculture, agro-food trade, and agrarian livelihoods in the Caribbean', in: David Goodman and Michael Watts (eds) *Globalising food: agrarian questions and global restructuring*, London: Routledge, pp. 119–32.

Raynolds, Laura T., Myhre, David, McMichael, Philip, Carro-Figueroa, Viviana and Buttel, Frederick H. (1993) 'The "new" internationalization of agriculture: a reformulation', *World Development* 21(7): 1101–1121.

Reardon, Thomas (1997) 'Using evidence of household income diversification to inform study of the rural nonfarm labor market in Africa', *World Development* 25(5): 735–47.

Redclift, Nanneke and Whatmore, Sarah (1990) 'Household, consumption and livelihood: ideologies and issues in rural research', in: Terry Marsden, Philip Lowe and Sarah Whatmore (eds) *Rural restructuring: global processes and their responses*, London: David Fulton, pp. 182–97.

Reichert, Christoph (1993) 'Labour migration and rural development in Egypt: a study of return migration in six villages', *Sociologia Ruralis* 33(1): 42–60.

Reid, Anthony (1988) *Southeast Asia in the age of commerce 1450–1680: Volume I – the lands below the winds*, New Haven, Ct: Yale University Press.

Reid, Anthony (1993) *Southeast Asia in the age of commerce 1450–1680: Volume II – Expansion and Crisis*, New Haven, Ct: Yale University Press.

Ricklefs, M.C. (1981) *A history of modern Indonesia*, Basingstoke: Macmillan.

Rigg, Jonathan (1985) 'The role of the environment in limiting the adoption of new rice technology in Northeastern Thailand', *Transactions of the Institute of British Geographers*, 10: 481–94.

Rigg, Jonathan (1986) 'Innovation and intensification in Northeastern Thailand: Brookfield applied', *Pacific Viewpoint* 27: 29–45.

Rigg, Jonathan (1989) *International contract labor migration and the village economy: the case of Tambon Don Han, Northeastern Thailand*, papers of East–West Population Institute no. 112, Honolulu: East–West Center.

Rigg, Jonathan (1994) 'Redefining the village and rural life: lessons from Southeast Asia', *Geographical Journal*, 160(2): 123–35.

Rigg, Jonathan (1997) *Southeast Asia: the human landscape of modernization and development*, London: Routledge.

Rigg, Jonathan (1998a) 'Rural–urban interactions, agriculture and wealth: a Southeast Asian perspective', *Progress in Human Geography*, 22(4) 1998: 497–522.

Rigg, Jonathan (1998b) 'Tracking the poor: the making of wealth and poverty in Thailand (1982–1994)', *International Journal of Social Economics* 25(6–8): 1128–1141.

Rigg, Jonathan (forthcoming(a)) 'Evolving rural–urban relations: changing patterns of life and livelihood' in Chia Lin Sien (ed.) *Southeast Asia transformed: a geography of change*, Singapore: Institute of Southeast Asian Studies.

Rigg, Jonathan (forthcoming(b)) 'Rural areas, rural people and the Asian crisis: ordinary people in a globalising world', in: Pietro Paolo Masina (ed.) *Rethinking development in East Asia: from illusory miracle to economic crisis*, Copenhagen: NIAS.

Rigg, Jonathan, Allott, Anna, Harrison, Rachel and Kratz, Ulrich (1999) 'Understanding languages of modernization: a Southeast Asian view', *Modern Asian Studies* 33(3): 581–602.

Ritchie, Mark A. (1993) 'The "village" in context: arenas of social action and historical change in Northern Thai peasant classes', paper presented at the 5th International Thai Studies Conference, SOAS, London, July 1993.

Ritchie, Mark A. (1996a) 'Centralization and diversification: from local to non-local economic reproduction and resource control in Northern Thailand', paper presented at the 6th International Conference on Thai Studies, Chiang Mai, Thailand, 14–17 October.

Ritchie, Mark A. (1996b) *From peasant farmers to construction workers: the breaking down of the boundaries between agrarian and urban life in Northern Thailand, 1974–1992*, Ann Arbor, Mich: UMI.

Roberts, Kenneth D. (1997) 'China's "tidal wave" of migrant labor: what can we learn from Mexican undocumented migration to the United States?', *International Migration Review* 31(2): 249–93.

Roberts, Rebecca (1996) 'Recasting the "agrarian question": the reproduction of family farming in the Southern High Plains', *Economic Geography* 72: 398–415.

Rodenburg, Janet (1997) *In the shadow of migration: rural women and their households in North Tapanuli. Indonesia*, Leiden: KITLV Press.

Roseberry, W. (1993) 'Beyond the agrarian question in Latin America', in F. Cooper, A. Isaacman, F. Mallon, W. Roseberry and S. Stern (eds) *Confronting historical paradigms*, Madison, Wis: University of Wisconsin Press.

Rotgé, Vincent L. (1992) 'Rural employment shift in the context of growing rural–urban linkages: trends and prospects for DIY [Daerah Istimewa Yogyakarta]', paper presented at the International Conference on Geography in the ASEAN Region, Gadjah Mada University, Yogyakarta, 31 August–3 September.

Russell, Margo (1993) 'Are households universal? On misunderstanding domestic groups in Swaziland', *Development and Change* 24(4): 755–85.

Saith, Ashwani (1991) 'Asian rural industrialization: context, features, strategies', in: Jan Breman and Sudipto Mundle (eds) *Rural transformation in Asia*, Delhi: Oxford University Press, pp. 458–89.

Sandee, Henry and Rietveld, Piet (1994) 'Promoting small scale and cottage industries in Indonesia: an impact analysis for Central Java', *Bulletin of Indonesian Economic Studies* 30(3): 115–42.

Sanderson, Steven E. (1986) *The transformation of Mexican agriculture: international structure and the politics of rural change*, Princeton, NJ: Princeton University Press.

Sanitsuda Ekachai (1990) *Behind the smile: voices of Thailand*, Bangkok: Post Publishing.

Sayer, Andrew (1989) 'Postfordism in question', *International Journal of Urban and Regional Research* 13(4): 666–95.

Scott, A.J. (1984) 'Industrial organization and the logic of the intra-metropolitan location, III: a case study of the women's dress industry in the greater Los Angeles region', *Economic Geography* 60(1): 3–27.

Scott, James C (1976) *The moral economy of the peasant: rebellion and subsistence in Southeast Asia*, New Haven, Ct: Yale University Press.

Scott, James C. (1985) *Weapons of the weak: everyday forms of peasant resistance*, New Haven, Ct: Yale University Press.

Shamsul, A.B. (1989) *Village: the imposed social construct in Malaysia's developmental initiatives*, Working paper no. 115, Sociology of Development Research Centre, Bielefeld: University of Bielefeld.

Shanin, Teodor (1987) 'Introduction: peasantry as a concept', in: Teodor Shanin (ed.) *Peasants and peasant societies: selected readings*, Oxford: Basil Blackwell, pp. 1–11.

Shanin, Teodor (1990) *Defining peasants: essays concerning rural societies, expolary economies, and learning from them in the contemporary world*, Oxford: Basil Blackwell.

Shanin, Teodor and Alavi, Hamza (1990) 'Peasants and capitalism: Karl Kautsky on "The agrarian question"', in: Teodor Shanin (ed.) *Defining peasants: essays concerning rural societies, expolary economies, and learning from them in the contemporary world*, Oxford: Basil Blackwell, pp. 251–79.

Sherman, D. George (1990) *Rice, rupees and ritual: economy and society among the Samosir Batak of Sumatra*, Stanford, Ca: Stanford University Press.

Sicular, Daniel T. (1991) 'Pockets of peasants in Indonesian cities: the case of scavengers', *World Development* 19(2/3): 137–61.

Singhanetra-Renard, Anchalee (1999) 'Population mobility and the transformation of the village community in Northern Thailand', *Asia Pacific Viewpoint* 40(1): 69–87.

Sjahrir, Kartini (1993) 'The informality of employment in construction: the case of Jakarta', in: Chris Manning and Joan Hardjono (eds) *Indonesia assessment 1993 – Labour: sharing in the benefits of growth?*, Political and Social Change Monograph no. 20, Research School of Pacific Studies, Canberra: Australian National University, pp. 240–58.

Smit, Warren (1998) 'The rural linkages of urban households in Durban, South Africa', *Environment and Urbanization* 10(1): 77–87.

Smyth, Russell (1998) 'Recent developments in rural enterprise reform in China: achievements, problems and prospects', *Asian Survey* 38(8): 784–800.

Svensson, Thommy (1991) 'Contractions and expansions: agrarian change in Java since 1830', in: Magnus Mörner and Thommy Svensson (eds) *The transformation of rural society in the Third World*, London: Routledge, pp. 145–85.

Tacoli, Cecilia (1996) 'Migrating "for the sake of the family"? Gender, life course and intra household relations among Filipino migrants in Rome', *Philippine Sociological Review* 44(1–4): 12–32.

Tambunan, Tulus (1995) 'Forces behind the growth of rural industries in developing countries: a survey of the literature and a case study from Indonesia', *Journal of Rural Studies* 11(2): 203–15.

Tana, Li (1996) *Peasants on the move: rural–urban migration in the Hanoi region*, Singapore: Institute of Southeast Asian Studies.

Tanabe, Shigeharu (1994) *Ecology and practical technology: peasant farming systems in Thailand*, Bangkok: White Lotus.

Tawney, R.H. (1932) *Land and labour in China*, London: George Allen and Unwin Ltd.

TDN (1997) 'IFCT sees poor economic growth', *Thai Development Newsletter* 33 (July–December), pp. 12–13.

Terweil, B.J. (1989) *Through travellers' eyes: an approach to early nineteenth century Thai history*, Bangkok: Duang Kamol.

Tomosugi, Takashi (1995) *Changing features of a rice-growing village in Central Thailand: a fixed-point study from 1967 to 1993*, Tokyo: the Centre for East Asian Cultural Studies.

Toyota, Mika (1999) 'Cross border mobility and multiple identity choices: the urban Akha in Chiang Mai, Thailand', PhD thesis, University of Hull.

Trankell, Ing-Britt (1993) *On the road in Laos: an anthropological study of road construction and rural communities*, Uppsala Research Reports in Cultural Anthropology No. 12, Uppsala: Uppsala University.

Tsubouchi, Yoshihiro (1995) 'A Malay village in Kelantan, 1970–1991', *Tonan Ajia Kenky (Southeast Asian Studies)* 33(3): 3–20.

Ulack, Richard (1983) 'Migration and intra-urban mobility: characteristics of squatters and urban dwellers', *Crossroads* 1(1): 49–59.

UNDP (1998) *Human development report 1998*, New York: Oxford University Press.

UNDP (1999) *Human development report 1999*, New York: Oxford University Press.

Ungpakorn, Ji Giles (1999) 'Thai workers in the 1990s: victims of agents of change?', paper presented at the 7th International Conference on Thai Studies, University of Amsterdam, Amsterdam, 4–8 July 1999.

van de Walle, Dominique (1996) *Infrastructure and poverty in Viet Nam*, LSMS Working Paper no. 121, Washington DC: World Bank.

Vandergeest, Peter (1991) 'Gifts and rights: cautionary notes on community self-help in Thailand', *Development and Change* 22: 421–43.

Vasana Chinvarakorn (1998) 'Collective wisdom', *Bangkok Post*, 21 March, 1998, http://www.bkkpost.samart.co/th/news/BParchive/BP19980321/210398_Outlook01.html.

Vickers, Adrian (1989) *Bali: a paradise created*, Berkeley, Ca: Periplus.

Vickers, Adrian (1996) 'Modernity and being *moderen*: an introduction', in: Adrian Vickers (ed.) *Being modern in Bali: image and change*, Monograph 43, Yale Southeast Asia Studies, New Haven, Ct pp. 1–37.

Vickers, Adrian (1997) '"Malay identity": modernity, invented tradition, and forms of knowledge', *Review of Indonesian and Malaysian Affairs (RIMA)* 31(1): 173–211.

Wade, Robert (1983) 'South Korea's agricultural development: the myth of the passive state', *Pacific Viewpoint* 24(1): 11–28.

Wang, Gabe T. and Xiaobo Hu (1999) 'Small town development and rural urbanization in China', *Journal of Contemporary Asia* 29(1): 76–94.

Wang, Mark Y.L. (1997a) 'Urban growth and the transformation of rural China: the case of southern Manchuria', *Asia Pacific Viewpoint* 38(1): 1–18.

Wang, Mark Y.L. (1997b) 'The disappearing rural-urban boundary: rural transformation in the Shenyang-Dalian region of China', *Third World Planning Review* 19(3): 229–50.

Ward, Neil (1995) 'Technological change and the regulation of pollution from agricultural pesticides', *Geoforum* 26(1): 19–33.

Ward, Neil and Almås, Reidar (1997) 'Explaining change in the international agro-food system', *Review of International Political Economy* 4(4): 611–29.

Watts, Michael (1994a) 'Life under contract: contract farming, agrarian restructuring, and flexible accumulation', in: Peter D. Little and Michael J. Watts (eds) *Living under contract: contract farming and agrarian transformation in sub-Saharan Africa*, Madison, Wis: University of Wisconsin Press, pp. 21–77.

Watts, Michael (1994b) 'Epilogue: contracting, social labor, and agrarian transitions', in: Peter D. Little and Michael J. Watts (eds) *Living under contract: contract farming and agrarian transformation in sub-Saharan Africa*, Madison, Wis: University of Wisconsin Press, pp. 248–57.

Watts, Michael and Goodman, David (1997) 'Agrarian questions: global appetite, local metabolism – nature, culture, and industry in *fin-de-siècle* agro-food systems', in: David Goodman and Michael Watts (eds) *Globalising food: agrarian questions and global restructuring*, London: Routledge, pp. 1–32.

Weixing Chen (1998) 'The political economy of rural industrialization in China: village conglomerates in Shandong province', *Modern China* 24(1): 73–96.

Whatmore, Sarah (1994) 'Global agro-food complexes and the refashioning of rural Europe', in: Ash Amin and Nigel Thrift (eds) *Globalization, institutions and regional development in Europe*, Oxford: Oxford University Press, pp. 46–67.

White, Benjamin (1976) 'Population, employment and involution in a Javanese village', *Development and Change* 7: 267–90.

White, Benjamin (1979) 'Political aspects of poverty, income distribution and their measurement: some examples from rural Java', *Development and Change* 10: 91–114.

White, Benjamin (1983) '"Agricultural involution" and its critics: twenty years on', *Bulletin of Concerned Asian Scholars* 15(2): 18–31.

White, Benjamin (1989) 'Problems in the empirical analysis of agrarian differentiation', in: Gillian Hart, Andrew Turton and Benjamin White (eds) *Agrarian transformations: local processes and the state in Southeast Asia*, Berkeley, Ca: University of California Press, pp. 15–30.

White, Benjamin (1991) 'Economic diversification and agrarian change in rural Java, 1900–1990', in: Paul Alexander, Peter Boomgaard and Ben White (eds) *In the shadow of agriculture: non-farm activities in the Javanese economy, past and present*, Amsterdam: Royal Tropical Institute, pp. 41–69.

White, Benjamin (1993) 'Industrial workers on West Java's urban fringe', in: Chris Manning and Joan Hardjono (eds) *Indonesia assessment 1993 – Labour: sharing in the benefits of growth?*, Political and Social Change Monograph no. 20, Research School of Pacific Studies, Canberra: Australian National University, pp. 127–38.

White, Benjamin and Wiradi, Gunawan (1989) 'Agrarian and nonagrarian bases of inequality in nine Javanese villages', in: Gillian Hart, Andrew Turton and Benjamin White (eds) *Agrarian transformations: local processes and the state in Southeast Asia*, Berkeley, Ca: University of California Press, pp. 266–302.

Wolf, Diane Lauren (1990) 'Daughters, decisions and domination: an empirical and conceptual critique of household strategies', *Development and Change* 21: 43–74.

Wolf, Diane Lauren (1992) *Factory daughters: gender, household dynamics, and rural industrialization in Java*, Berkeley, Ca: University of California Press.

Wolf, Eric R. (1967) 'Closed corporate peasant communities in Mesoamerica and Central Java', in: Jack M. Potter, May N. Diaz and George M. Foster (eds) *Peasant society: a reader*, Boston, Ma: Little, Brown and Company, pp. 230–46 (first published in *South Western Journal of Anthropology* 13(1): 1–18, 1957).

Wolf, Eric R. (1986) 'The vicissitudes of the closed corporate peasant community', *American Ethnologist* 13(2): 325–29.

Wolpe, Harold (1972) 'Capitalism and cheap labour-power in South Africa: from segregation to apartheid', *Economy and Society* 1: 425–56.

Wong, Diana (1987) *Peasants in the making: Malaysia's green revolution*, Singapore: Institute of Southeast Asian Studies.

World Bank (1993) *World development report 1993*, New York: Oxford University Press.

World Bank (1997) *World development report 1997*, New York: Oxford University Press.

World Bank (1998) 'Social consequences of the East Asian financial crisis', http://www.worldbank.org/poverty/eacrisis/partners/library/socconsq/index.html.

World Bank (2000) *World development report 1999/2000*, Oxford: Oxford University Press.

Xenos, Peter and Kabamalan, Midea (1998a) 'The changing demographic and social profile of youth in Asia', *Asia–Pacific Research Reports no. 12*, Hawaii: East–West Center Program on Population.

Xenos, Peter with Kabamalan, Midea (1998b) *The social demography of Asian youth: a reconstruction over 1950–1990 and projections to 2025*, East–West Center Working Papers no. 102 (May), Honolulu: East–West Center.

Xenos, Peter, Kabamalan, Midea and Westley, Sidney B. (1999) 'A look at Asia's changing youth population', *Asia Pacific Population and Policy* 48 (January).

Young, Alwyn (1995) 'The tyranny of numbers: confronting the statistical realities of the East Asian growth experience', *Quarterly Journal of Economics* 110(3): 641–80.

Zaman, Habiba (1995) 'Patterns of activity and use of time in rural Bangladesh: class, gender and seasonal variations', *Journal of Developing Areas* 29: 371–88.

Zoomers, Annelies E.B. and Kleinpenning, Jan (1996) 'Livelihood and urban-rural relations in Central Paraguay', *Tijdschrift voor Economische en Sociale Geografie* 87(2): 161–74.

Index

Information in footnotes, that is not included in the main text on the same page, is indexed in the form 28(n1), ie note 1 on page 28

Printed and bound by CPI Group (UK) Ltd, Croydon, CR0 4YY

01/11/2024

01782615-0020